Human Resources Management in the Hospitality Industry

Human Resources Management in the Hospitality Industry

Editor

Delphina Fonseca

scitus
academics

Human Resources Management in the Hospitality Industry
Edited by **Delphina Fonseca**

ISBN: 978-1-68117-547-8
Library of Congress Control Number: 2016942951

www.scitusacademics.com
Box No. 4766, 616 Corporate Way,
Suite 2, Valley Cottage,
NY 10989

Printed in United States of America on Acid Free Paper

Preface

To have maintained its place as an essential text for so many years Human Resource Management in the Hospitality Industry has had to remain relevant to its broad audience. Today's hospitality professional must be an expert at managing many functions. Altogether, human resource management is the process of proper and maximizes utilization of available limited skilled workforce. The core purpose of the human resource management is to make efficient use of existing human resource in the organization. Managing the workforce effectively is perhaps the single most important issue in delivering first-class service which underpins a hotel's performance, reputation and profitability. At the international level, the countries of origin of hotel staff, particularly management and supervisors, have changed dramatically. In the past, companies, in order to preserve their brand integrity in new countries, tended to post skilled staff from their own already successful properties. In many countries, however, immigration rules have changed in order to protect local employment, ensuring that home-grown employees have employment priority. In addition, more international companies can now rely on a more abundant source of local trained labor in domestic markets, as a result of local education and training initiatives. Even so, restrictions on importing senior management and skilled supervisors present companies with challenges as they work to protect the integrity of their brand.

Human Resource Management in the Hospitality Industry provides the reader with an essential understanding of the purpose, policies and processes concerned with managing an enterprise's workforce within the current business and social environment. This book examines the nature and assesses the impact of HRM within a highly under-researched division of the service sector, namely the hospitality industry. The book provides comprehensive overview of the hospitality and tourism industries, and emphasizes the importance of effective human resources management to all organizations within them.

Table of Contents

Chapter 1 Human Resource Management Ethics and Professionals' Dilemmas: A Review and Research Agenda 1

Chapter 2 Strategic Human Resource Management in U.S. Luxury Resorts—A Case Study 25

Chapter 3 Historical Evolution of Human Resource Information System (HRIS): An Interface between HR and Computer Technology 41

Chapter 4 Why Should Gender Differences In Hospitality Really Matter? A Study of Personnel's Service Orientation and Job Satisfaction in Hotels 55

Chapter 5 On Becoming a Strategic Partner: The Role of Human Resources in Gaining Competitive Advantage 83

Chapter 6 Key Issues in Strategic Human Resources 109

Chapter 7 Preparing Hospitality Organizations for Self-Service Technology 139

Chapter 8 Recruitment and Selection Process and Employee Competence Outcome: An Important Area for Future Research 159

Chapter 9 A Case Study into the Benefits of Management Training Programs: Impacts on Hotel Employee Turnover and Satisfaction Level 179

Chapter 10 How Individual Beliefs Impact Individual Performance 195

Chapter 11 Explaining Intrapreneurial Behaviors of Employees with Perceived Organizational Climate and Testing the Mediating Role of Organizational Identification: A Research Study among Employees of Turkish Innovative Firms 223

Chapter 12 Employee Compensation: Theory, Practice, and Evidence 241

Chapter 13 Role of Job Satisfaction in the relationship of Business Excellence and OCB: Iranian Hospitality Industry 269

Chapter 14 Motivating Your Staff to Provide Outstanding Service 285

Index 301

Chapter 1

Human Resource Management Ethics and Professionals' Dilemmas: A Review and Research Agenda

Kelechi John Ekuma, Loliya Agbani Akobo

Institute for Development Policy and Management, University of Manchester, Manchester, M13 9PL, UK

ABSTRACT

The global nature of competition, new and ever changing employee expectations, changing societal values and constant revisions of employment law have propelled human resource management (HRM) as one of the critical business function for continued organisational competitiveness in contemporary times. This increased importance of HRM has arguably added new set of responsibilities with wide-ranging implications for HR professionals. In specific terms, the emergence of HRM as a panacea for integrating business strategy and people management has exposed personnel practitioners to a new set of role demands, professional challenges and management expectations. The ethical implications of these demands and expectations are significant and needs to be critically analysed. But whilst there has been an increasing interest generally in business ethics amongst scholars and practitioners, the employment of people in contemporary times, gives rise to unique and critical ethical dilemmas on the part of HR practitioners which have been largely ignored. This is a significant gap in the literature which needs to be addressed. This paper makes a contribution towards addressing this gap. Synthesising literatures in HRM, philosophy and management studies, the paper critically examines frameworks, contemporary debates and emerging themes around HR ethics, raising a number of fundamental issues that should be addressed by scholars, in line with the changing realities of contemporary business environment, particularly the management of people. The main aim of this analysis is to highlight some of the emerging or sometimes ignored aspects of HR ethics with a view of situating a future research agenda to help identify appropriate conceptual lenses and realistic strategies for addressing issues of practitioner dilemmas in contemporary people management.

Keywords: Ethics, Human resource management, Business ethics, Ethical framework, Ethical climate

1. INTRODUCTION

So much has been written about the role of the Human Resource Management (HRM) function and its ability to respond to competitive pressures (Foote, 2001). In recent times, notions of 'best practice' HRM exemplified in the works of Huselid (1995), Guest (2001) and several others, have emphasised the importance of people as a source of competitive advantage and advocated the use of combinations of specific HRM practices to achieve this. Within such approaches is an implicit assumption that gaining employee commitment to the organisation's goals is possible and indeed crucial (Foote, 2001).

There is evidence, however, of a gap between the rhetoric and reality of HRM in that many organisations are eroding job security, while at the same time increasing the demands made of employees (Foote, 2001). It is this gap that has led researchers such as Legge (1995) and Purcell and Hutchinson (1996) for instance, to postulate that restructuring initiatives and work intensification have contributed to employee demoralisation, de-motivation and alienation, instead of enhanced engagement thereof and raises critical ethical concerns.

In determining the ethical course of action, HR professionals have to consider their dual memberships in the business organization and in the profession. Their professional loyalties may place them in direct conflict with their organization's business goals (Wiley, 1998: 147). It is not unusual for the HR practitioner's values and expectations to be contrary to the organization's values (Archer, 1986: 97-98, cited in Wiley 1998: 147). Because professionals place a premium on expertise, specialization and objectivity, they are appalled when they observe managers making decisions based upon non-objective techniques, intuitive speculation or seat-of-the-pants approaches (Wiley, 1998).

Given the foregoing, the salient question that then follows is: how can HR reconcile the need for professionalism and current propensities and pressures for organisational profit maximization on the one hand, and the need to promote positive values, adhere to ethical standards and enhance employees' welfare, expectations and general engagement on the other? Admittedly, there is no one right approach to illuminating and understanding this question. Yet, as Wiley (1999: 157) observes, this does not mean that nothing should be done, as research in this area is necessary and researchers and other interested parties must continue to explore the perceptions and practices among HR managers to better understand and learn from them. Indeed, there has been an increasing interest generally in business ethics, particularly in the U.S. and increasingly in the U.K. (e.g. Winstanley & Woodhall, 2000), however, the employment of people particularly in contemporary times, gives rise to unique and critical ethical dilemmas on the part of HR practitioners which have been largely ignored.

Through the integration of literatures in human resources, management studies and philosophy, this paper seeks to direct the attention of scholars and practitioners alike to these ignored aspects of HR ethics. In doing this, the paper critically examines contemporary debates and emerging themes around HR ethics, raising a number of fundamental ethical questions which current literatures alluded to but failed to address in line with the changing realities of contemporary business environment, particularly the management of people. This analysis is intended to help explain and elicit further debates on the many shortcomings inherent in contemporary thinking about HR ethics. In this regard therefore, the overall aim of the article is to provide a research agenda to identifying realistic strategies and conceptual lenses for addressing issues of HR ethics in contemporary people management.

2. CONCEPTUALIZING ETHICS

The theory of ethics can be extremely complex, and in order to be of any use in day – to – day work contexts, it must be made practical (Kew & Stradwick, 2008). It is this critical need that presents the immediate major challenge to management scholars and Human Resource (HR) practitioners and engenders most of the theory – practice gap. As Stewart and Rigg (2011) point out, despite the long-standing effort at theorising and understanding the concept of ethics and making it practical, there is relatively little agreement or certainties, even as there are many different perspectives on meanings and applications.

Indeed, the concept has defied a univocal definition; neither is there any single understanding. Fisher and Lovell (2009) attribute this development partly due to the fact that there are very few studies of ethics built on theoretical positions, but are largely based on beliefs and arguments. It is this line of reasoning that perhaps, informs Stewart and Rigg (2011: 287) argument that 'ethics cannot be taught; rather it is incumbent on individuals to decipher ethical positions for themselves'. What can be taught in Stewart and Rigg's (2011) opinion is a range of ideas and tools that can help with that deciphering or working out process.

This development notwithstanding, some scholars have attempted to provide a better understanding of the concept. Legge (2007: 39) for instance, presents a very succinct definition of the concept as 'the identification of the good and its just or fair distribution'. This view is consistent with that of Stewart and Rigg (2011: 288) who see ethics as 'the study of right and wrong as well as to a specification of what is right and what is wrong'. Torrington and Hall (1998: 682) cited in Stewart and Rigg (2011: 288) suggest an important distinction between the singular and plural forms of the word, arguing that the singular refers to 'moral value' and the 'principles that ought to govern conduct'; while the plural describes 'codes of behaviour considered to be correct, especially that of a particular group or profession'. A broader definition of the concept is presented by Cornock and Johns (1995) to include:

- Fairness.
- Deciding what is right and wrong.
- The practices and rules that underpin responsible conduct between individuals and groups.
- Billington (2003: 20-25) cited in Kew and Stradwick (2008: 274) lists some of the distinctive characteristics of ethics to include:
- Nobody can avoid ethical decisions. We all make ethical decisions every day.
- Other people are always involved in ethical decisions. There is no such thing as private morality.
- Ethical decisions matter – they affect the lives of others.
- Although ethics is about right and wrong, there are no definitive answers. The philosopher can forward principles which should guide decisions, but the ultimate decision is always down to the individual.
- Ethics is always about choice – a decision where the individual has no choice cannot be considered unethical.

Beauchamp and Bowie (1997) further note that ethics involves an inquiry into the justification or rationale for those standards. While by no means exhaustive, the foregoing definitions emphasise the importance of morals, fairness and the centrality of the individual in the practicality of the concept of ethics. Billington (2003) corroborates this position, when arguing that ethics entails the values that a person seeks to express in a certain situation and the morals that direct how he/she intends to achieve this. Morality (although contested) has been defined as 'a social institution composed of a set of standards pervasively acknowledged by members of a culture' (Beauchamp & Bowie, 1997:1).

Given the foregoing, it is plausible to argue that while it is true that there is no consensus on definition, there seems to be a general agreement that ethics is concerned with 'establishing the principles of what is right or moral and subsequently transmitting what is established as right or moral into a system of codes or rules which govern behaviour of individuals in a particular context (Stewart & Rigg, 2011). What is important therefore is to be clear about what is being established and about which meaning is being applied to it. Indeed, the need to be clear about what is being established and the meaning applied thereof becomes even more critical, considering the eclectic and contested nature of the concept of ethics.

3. THE NATURE OF BUSINESS AND HR ETHICS

All The application of the study of ethics to business, particularly to organisation and management theory, is a relatively recent development (Stewart & Rigg, 2011). Winstanley and Woodhall (2000) and Torrington and Hall (1998) in their various works, suggest that ethical issues have been of marginal significance to the unfolding academic debates around management practice, particularly HRM and that the concept can be for many people, 'incongruous' in the context of business organisations. However, as Palmer (2007) notes, theorists from different backgrounds, have pondered the ethical basis of businesses and employment relations for decades. Fisher and Lovell (2009) cited in Stewart and Rigg (2011: 287) present a similar perspective when they argued that while it is true that applying the notion of ethics to organisations, businesses and otherwise is relatively recent, the notion has been the focus of human enquiry for centuries.

In a contemporary context, the increasing importance of Corporate Social Responsibility (CSR) associated with publicised scandals such as the well-known Enron case has led to a sustained interest in and more attention to the concept of ethics in business and management practice and research (Stewart & Rigg, 2011). Consequently, business ethics as a field of study and as an issue with currency in broader community has grown considerably in recent times.

Recent research on business ethics indicate that HR-related issues are increasingly prominent in the day –to- day practice of contemporary managers (Wooten, 2001). As Winstanley, Woodall & Heery (1996) observe, the relationship between ethics and human resource management is emerging as a subject of serious academic enquiry. While some (e.g. Welch, 1993) have argued that general organisational ethics and ethics of HRM practice are not indistinguishable, a contrary perspective can nonetheless be presented. In line with Wooten's (2001) argument, the relative circumstances and ethical challenges of HRM professionals who develop and maintain distinctive competencies and engage in highly specialised functions and duties, must be unique in several respects to those of other management practitioners.

Admittedly, many organisational members engage in HRM activities (e.g. training, selection, appraisal and so on) as a function of their job roles, however, the fact remains that they are often never required nor expected to have the same level of competence as an HRM professional nor are they held to the same standard of practice (Wooten, 2001). Given the above, it is plausible to conclude therefore, that the ethics of HRM while sharing many features with general organisational ethics must at least be relatively unique in several respects.

That said, the debate on ethical issues in the employment relationship can be linked to extant debates in employment (Greenwood & De Cieri, 2007). In recent times, HRM has come under intense criticism as being very manipulative, exploitative and in fact immoral. Critical writers such as Townley (1993), Keenoy & Anthony (1992), Sennett (1999) and Guest (1998) have suggested in their various works that HRM practice objectifies individuals; suppresses resistance and confrontation and creates a new reality through its rhetoric and manipulative tendencies.

This argument is not unconnected with the fact that in a globalising world of increased emphasis on competitiveness and profitability, HR has emerged as a major panacea for effective management of people to achieve organisational goals and sustainable advantage. However as Greenwood (2002: 261) argues, the word 'management' in this context is a euphemism for 'use'. It is this that has led some (e.g. Stone, 1998: 4) to define HRM as 'the productive use of people in achieving the organisation's strategic business objectives and the satisfaction of individual employee needs'. As Greenwood (2002: 4) succinctly sums it:

Whether *hard* or *soft* (HR), or some combination of the two, it is only the nature and extent of this 'use' which varies; and HRM implicitly accepts such 'use'...it is also hazardous to justify HRM on the basis that it maximises positive outcomes for those affected. Who is to say what is good and who should be included?

Consistent with the perspective of the Theory 'X' (McGregor, 1960) in the context of HRM, there is a popular image of the profit motivated HR manager – just like any other manager – whose primary concern is the reduction of cost of production and profit maximization, and for whom it has been argued, ethics is a veneer to cloak greed. In this view and in the context of HR ethics, the manager is seen as amoral and does not find ethical considerations central to his decisions. For him, it would appear that winning is the only thing that matters and maximizing people's potentials to win is the only important priority (Danley et al, 1996). While this might not necessarily represent the contemporary HR manager's ethical stand (who is guided and bound by several professional and work ethics), the mantra of maximizing worker's potentials, enhancing efficiency and achieving strategic objectives, makes it difficult for HR to totally disassociate itself from these allegations.

Whatever the case, what is clear and should not elicit much debate is that HRM maximizes employees' potentials and helps in achieving strategic business objectives, through a combination of different HR strategies and policies. What is important therefore, and of primary concern to this study, is how HR delivers on these important functions, considering the many ethical challenges that it presents in a contemporary context. Herein lay the importance of ethical consideration in HRM and the linkages thereof.

An assessment of several works on Business and HR ethics, indicate that much of the literature argues explicitly or implicitly that HRM as an approach to management, with its prescriptions of labour flexibility, unitarism, individualisation, and business strategy integration, requires ethical evaluation and critical analysis (Cornelius & Gagnon, 1999: 226). A developing literature on the ethical dimensions of HRM has begun to critically examine these issues; and while some (e.g. Danley et al, 1996) have presented a fairly positive assessment of the situation; others (e.g. Greenwood, 2002) have not been that very complimentary.

Winstanley & Woodhall (2000a) in their edited compilation summarise several recent debates about ethics and HRM. They briefly highlighted how various management theories might been seen to approach ethics and HRM, and then explore some ethical dilemmas associated with particular HRM practices, such as recruitment and selection, training and development, work practices, remuneration and employee participation. They were quick to observe that practitioners face tremendous dilemmas in the course of their work and concluded that there are significant constraints facing ethical HRM (Palmer, 2007). What they however failed to do is provide a panacea or at least a framework that might help mitigate these identified constraints.

In the same vein, Pinnington, Macklin & Campbell (2007) in their own edited compilation: *Human Resource Management: Ethics and Employment,* similarly examine ethics and employment issues in contemporary Human Resource Management. They note that bringing ethical awareness into the core of HRM is all the more important given the trend in Western societies towards the decline of trade unionism and the emergence of more individualist approaches to employment, arguing that the decline of collectivist arrangements has left many employees potentially more vulnerable to opportunistic and unethical behavior. While this position is relatively valid, it pertinent to note that the decline of unionism or the increasing individualist approach to employment, does not in themselves lead to an increase in unethical behaviour. Rather, it could be argued that increasing global competition and the need to cut costs are more valid sources of unethical behaviour in our view.

Overall, one of the most highly evident themes in both Winstanley & Woodhall's (2000a) and Pinnington et al., (2007) works is the ideological tension between individualism and collectivism and especially the increasing vulnerability of many employees when trade union protection is reduced while the collective power of the corporation is enhanced. It is this that has led Cornelius and Gagnon (1999) to argue that the increasing focus on HR ethics may be seen as filling a vacuum left by the declining presence and power of trade union in British and indeed, other societies' employment relations.

As Pinnington et al., (2007: 19) rightly note, while this may benefit economic performance and may be justified in terms of the general well-being, it has some stark and, for some, unacceptable consequences for those whose economic security is at the mercy of market imperatives. Given the above, the fundamental question is: can and should HRM simply seek to mitigate these consequences in individual cases, or could there be a more positive and systematic approach to the CSR of companies to their employee stakeholders?

Against the backdrop of this seeming ambivalence, Mathis and Jackson (1997) have argued that ethical issues in HRM pose fundamental questions about fairness, justice, truthfulness, and social responsibility. They note that seven specific ethical issues in HRM have the potentiality of creating particular difficulty. They further identified these as: withholding information on a problem employee with another potential employer, investigating credit and criminal records of potential employees, obligations to long-term employees who become ineffective due to job changes, considerations involving lifestyle upon professional opportunities, enforceability of smoking cessation, employment considerations for applicants with dependents with costly health concerns, and privacy issues involving AIDS.

Current research (e.g. Appelbaum, Deguire & Lay, 2005; Desphande, 1996, etcetera) suggest that one important factor that may influence ethical behaviour of employees is the ethical climate of an organization. The ethical climate of an organization is defined by the shared perception of how ethical issues should be addressed and what constitutes an ethically correct behaviour (Desphande, 1996). Appelbaum rt al., (2005), are of the view that ethical climate directly influences the corporate culture of organisations, and in turn profoundly influences the disposition of HR practitioners to ethical issues in the workplace. The authors however failed to tell us what in turn influences the ethical climate in the first place. This would have provided an insight into the epistemological biases of HR professionals and how their work ethics are formed or established, considering the fact that all human behaviours have been shown to be influenced by epistemological basis.

In this regard, several scholarly attempts have since explored the epistemological considerations of HRM professionals. Dachler and Enderle (1989) cited in Wooten (2001) for instance, have examined a variety of the epistemological assumptions of those in the HRM field. They argue that the behavior of HRM professionals and their decisions reflect their implicit assumptions and values. Dachler and Enderle also argue that many ethical issues are ignored by HRM professionals as a result of these assumptions and values.

Payne (1994) has examined alternative social paradigms explaining HRM professional behavior, ranging from constructivist to postmodern approaches. Payne suggests that the use of multidimensional paradigms is greatly needed to influence 'firm' ethical awareness and judgment (Wooten, 2001). Similarly, Greenwood (2002) has also analyzed the conceptual basis of HR ethics, examining issues such as affirmative obligation, the concept of stake-holding and the dual nature of HRM in modern organizations in the context of issues of rights, obligation and interferences.

While these works generally acknowledge an increased and sustained awareness on the need for a more 'ethical' HRM, none however question the fact that HRM has the potentiality and capacity to treat employees instrumentally as a resource, properly so called. A common theme that nonetheless permeates is a broad unanimity on the fact HRM as practiced in modern organizations, has on the average, generally failed the ethical test. As Wooten (2001: 173) succinctly sums it:

While the field (of HRM) has grown in sheer numbers, as well as strategic importance to organizational functioning, it has lagged behind other professions such as accounting or law to the extent that there is specificity concerning core values of the profession, clear expectations concerning expected roles, standards of practice, and the processes to safeguarding the public from inappropriate professional behaviour. Greenwood (2002: 275) echoes a similar sentiment when he argues that:

...the majority of people would accept the ideas that an individual or organisation must treat individuals with respect, and that an individual or organisation does not have the right to interfere with the freedom of an individual. These two minimum standards are used to assess HRM. HRM in its various guises fails this evaluation.

In the light of this spectrum of ethical limitations on the part of HRM and without any claims of providing or projecting any 'best practice' paradigm, this paper suggests that there is a need for a critical re-evaluation of the ethical basis of the HR function and the evolution of principles which may be fairly equitable to all stakeholders in the employment relationship. This may need a critical re-evaluation of current mainstream ethical frameworks in the context of HRM. The next few paragraphs critically examine these issues.

4. ETHICAL FRAMEWORKS AND HRM

The ethical approaches to HRM are diverse, so one of the interests of this review is to depict different theoretical frameworks and ethical arguments for the management of human resources in organizations from a critical perspective. It is pertinent to point out that while several ethical paradigms can be applied in the analysis of HRM, for our purpose however, a multi-faced approach (see Winstanley & Woodall, 2000) is adopted based on the following categorisation: ethical theories of rights, obligations, distributive justice and ethics of care, from a critical perspective. This does not in any way suggest that other approaches are less important or have become obsolete, but because of its centrality to the HR function and the fact that advocates of the 'new deal' in employment have grounded their argument about the change in the psychological contract between employer and employee in some, if not all of these categorizations (Winstanley, et al., 1996).

Yet, contrary to popular conceptions and as Winstanley & Woodall (2000) note, these are not self-evident truths, but contentious issues for debate. It is thus important that human resource professionals and academics critically re-engage with some of these ethical issues in line with the exigencies of the moment. What is intended here is to situate them in relation to the HRM implications and in the context of our analysis and contemporary realities. This is in line with Winstanley & Woodhall's (2000: 8) argument that 'the professional and academic HR community tends to have a different understanding of what "ethical" concern means'.

4.1. Ethical Theories of Right and Obligation

As Winstanley and Woodall (2000: 11) note, right based ethical frameworks tends to draw on two key concepts from the philosopher – Immanuel Kant as follows:

a. The principle that that what is right for one person is right for everyone, and thus it is important to do unto others as you would be done by – the criteria of universality and reversibility.

b. The principle of respect for people whereby they should be treated as ends in themselves and never as means to an end.

This Kantian framework forms the basis of *deontological* perspective to business ethics because of its focus on duty, and its perspective that links ethics to things that are good in themselves (Abreu & Badii, 2006). It epitomizes what Winstanley and Woodall (2000: 11) describe as 'worth of the individual'. Abrue and Badii (2006: 110a) observe that the theory of Kant proposes several rights that concern the following issues:

1) The fundamental right to life and safety.
2) The human rights to privacy.
3) Freedom of conscience.
4) Freedom to speech
5) Freedom to private property.

In the context of HRM and as Winstanley and Woodall (2000: 12) rightly argue, rights-based frameworks continue to be relevant to HRM, particularly in areas such as:

- Selection interviewing (the right to privacy and confidentiality of personal information, particularly where it is not relevant to the job and the obligation on the part of the employer to ensure that).
- Occupational testing (such as the right to feedback from the organisation).
- Equal opportunities and diversity management (the right to be treated the same or to be given special treatment).
- Flexible employment contracts and working time (the right to 'family-friendly' practices).
- Whistleblowing (the right to speak out about wrongdoing).
- Staff charters (which may outline employee rights and responsibilities; as well as the employer's obligation to the employee).
- Employee development (the right to psychological and physical safety, such as with relation to outdoor training, for instance).

From the foregoing, it plausible to affirm that rights-based ethical framework and its expected obligations (both from employees and employers) will undoubtedly protect the interest of individual employees and guarantees a reasonably more ethical HRM. Yet, it has nonetheless come under severe criticisms from proponents of the*business case* school who maintain that it is impractical and undermines the interest of businesses; even as it does not generate much enthusiasm from HRM practitioners.

Against the backdrop of the need to promote the interest of the business, rights-based ethical frameworks has been contrasted by the 'business case' argument underpinned in the concept of Ethical egoism. Beauchamp and Bowie (1983: 18) cited in Winstanley and Woodall (2000: 10) argue that ethical egoism is based on the Hobbesian assumption that 'the only valid standard of conduct is the obligation to promote one's own well-being above anyone else's'.

It is often argued that an important factor to consider when thinking about the ethics of HRM is that of the aims and purpose of the employer. Christy and Brown (2009: 70) corroborate this assertion when they argue that the organizational purpose relates directly to ethics in HRM because organizations can be taken to employ people primarily in order to pursue the purpose of that organization. Accordingly and in the context of HRM, ethical egoists maintain that the ethical role of HR professionals would be limited to supporting the enlightened self-interest of the employer rather than the rights of employees (Winstanley and Woodall, 2000b: 10).

From an objective stand point, it would appear that the *business case* paradigm is a very commonly used ethical argument in HRM practice. As Winstanley, Woodall & Heery (1996) argue, the prevailing common-sense ethical framework seems to justify HRM policy in terms of its 'utility to the organization' or its 'consequences'. This is consistent with the classic view of the stakeholder theory (shareholder value theory) (see Friedman, 1970) and reflects much of the thinking that underpins the concept of ethical egoism. This perspective defines the sole purpose of business as that of maximizing shareholder value through profits in line with Friedman (1970) and Stenberg's (1994) argument that the sole purpose of an orginsation's social responsibility is to make as much profit as possible for the owners, as long as it is within the law and rules of the 'game'. This is akin to the utilitarian and consequentialist frameworks identified earlier in this review.

It is my contention here however, that utilitarianism, or consequentialism (or any of its variants) is a very weak principle for ethical action and therefore, should not profoundly influence or justify HR practices. The fact that an act does not harm people or constitutes the greatest 'good'; or because it maximizes business interests, does not make it ethical in itself. We can also adapt Winstanely, Woodall and Heery's (1996) argument and question who decides the overall organizational goals against which HRM policy is judged ethical.

Moreover, it is also logical to argue that the definition of the 'purpose of business' can be broadened to suggest as Christy and Brown (2009: 72) note, that a business has a wider responsibility to the society in which it operates and that the task of a manager must recognize additional duties associated with the needs and claims of various stakeholder groups.

While I recognize the many limitations and seeming conceptual confusion surrounding the concept of the stakeholder theory, it is this perspective that a 'business has a wider responsibility to the society', that informs his understanding of the concept in the context of this research. Therefore, I adopt Johnson and Scholes' (1997) cited in Kew and Stradwick (2008: 282) broader definition that stakeholders are 'those individuals or groups who depend on the organization to fulfill their own goals and whom, in turn, the organization depends', not necessarily those with legal or financial relationship with the organization as the CIPD (2003: 30) would suggest. This position is akin to what Winstanley and Woodall (2000b: 13) categorize as 'ethical humanism' and emphasizes the dignity and centrality of the individual in the employment relationship and also asserts his/her right as a major stakeholder.

4.2. Distributive Justice and Ethics of Care

In addition to the foregoing approaches, a number of alternative ethical frameworks also lend themselves to the analyses of HRM. Some of these approaches as identified by Winstanley, Woodall and Heery (1996: 9) include, but are not limited to:

1) *Social and organizational justice*. In the context of HRM, the procedural principles of distributive justice, egalitarianism, equity, equality of opportunity can be used, for example, to evaluate systems for pay setting, recruitment and performance management.
2) *Universalism*. This is very similar to the Kantian principle of treating each individual as an end in themselves and not just as a means to an end, or sacrificed in a utilitarian equation of the greatest good for the greatest number. Doing as we should wish to be done by is a very strong ethical principle.
3) *Community of purpose*. This approach moves from the individual to consider the roles and responsibilities of organizations. A more communitarian view of the organization, or at least taking a stakeholder rather than purely a shareholder view of the firm, stresses the need for debate, respect and tolerance as virtues to be held above all else, for example, when assessing strategies for downsizing and delayering.

In addition, Gilligan (1982, 1987) cited in Winstanley and Woodall (2000b: 14) has also introduced subjective and intuitive approaches to resolving ethical issues in the workplace, arguing that moral judgments need to be sensitive to both the needs of the situation and other individuals. Gilligan maintained that being impartial makes it difficult to imagine oneself in other's position and thus understand the other's perspective (Carse, 1996: 86, cited in Winstanley and Woodhall, 2000b: 15).

It is this line of reasoning that forms the basis of the concept of *ethics of care* which is grounded on the premise that moral reasoning involves empathy and concern, emphasizing responsiveness and responsibility in our relations with others, where moral choices are made in relationship with others, not in isolation (Gilligand, 1987). As Gilligand (1987: 24) aptly puts it:

As a framework for ethical decision, care is grounded in the assumption that detachment, whether from self or from others is morally problematic, since it breeds moral blindness or indifference – a failure to discern or respond to need.

Similarly, Cornelius and Gagnon (1999) have also attempted to introduce and legitimize emotional dimension to ethical reasoning in HRM through their concept of *emotional labour*, arguing that frameworks that rests primarily on intellectual reason, ignores the salient fact that people are likely to respond emotionally to challenges to deeply held ethical beliefs. As they (i.e. Cornelius & Gagnon) succinctly sum it:

We would like to suggest that the emotional dimension should be considered in developing frameworks for understanding ethics in action, through the concept of 'emotional labour'. We would argue that within the context of the ethical landscape should be incorporated the 'emotional landscape', personal and shred, as this is likely to inform the perceived legitimacy of ethical action (1999: 231).

The foregoing analysis on ethical frameworks and their relevance to HRM is by no means exhaustive, yet, it suffices to illuminate our understanding on the 'ethical' practice of HRM and the justification or otherwise of some of its contemporary policies. However, from a critical perspective, it would appear that there are problems with these arguments and frameworks with regards to its adoption by human resource management practitioners. In specific terms, the humanistic values of empathy, acceptance, genuineness, congruence and unconditional positive regard which underpins most of the argument for all the approaches examined above, including *ethics of care* and that of *emotional labour* are very different values from those that underpin contemporary HRM practice (Winstanley & Woodall, 2000b).

Admittedly, there is the need for HRM to become more 'ethical' in the execution of its functions, but this need does not in anyway, undermine the fact that the sole purpose of all HR functions, is to enhance the efficiency and increase the productivity of organisations through the maximisation of employees' potentials. If we accept this position, then it is logical to suggest that the *business case* argument will always almost prevail in HR practices and policies and no doubt increase the 'ethical' pressure on the practitioner.

5. CONTEMPORARY DEBATES AND EMERGING THEMES IN HRM ETHICS: SITUATING THE DISCOURSE ON PROFESSIONALS' DILEMMAS

Recent thinking involving ethical dilemmas in HRM has been very broad and reflects the programmatic and contemporary issues of HRM practice

(Wooten, 2001). For example, several authors have raised the issue of ethics in both drug testing (e.g. Birsch, 1995; Carson, 1995) and genetic screening (e.g. Kupfer, 1993). Others have also critically examined the ethical issues and dilemmas involved in the practice of total quality management (e.g. Fort, 1995; Raiborn & Payne, 1996). Further, the ethics inherent in the practice of outsourcing, tele-working and work autonomy have also recently received attention from scholars (e.g. Henderson, 1997; Brey, 1999; Taskin & Devos, 2005).

Similarly, in recent years, the ethics of remuneration and pay and benefits have also become a matter of keen interest as a result of the controversy over bogus executive pay and more recently, because of the cataclysmic consequences of the global economic recession (e.g. Heery, 1996; Stewart & Rigg, 2011). The ethics of learning and talent development has also engaged the interest of recent writers (e.g. Stewart & Rigg, 2011); so also is that of contemporary human resource development (e.g. Woodall & Douglas, 1999). Equally eliciting interest is the ethics of performance management and measurement and issues of individual privacy (e.g. Stuart- Smith, 1996). Following Winstanley et al., (1996: 6-8), these contemporary ethical issues can be summarized into four broad categories as follows:

i. **Insecurity and risk**: debate over the precise extent and significance of the trend towards greater job insecurity is ongoing but for many employees the world of work has become less secure in recent years. This has arisen partly as a result of changes in the macroeconomic climate and the re-emergence of mass unemployment; in part from business restructuring and the stripping out of costs through redundancy, delayering and outsourcing; and also it derives from the adoption of more contingent contracts of employment and systems of reward.

ii. **ii. Surveillance and control**: this is the consequence of the spread of new forms of work organization and management control. These range from the use of psychometric tests for recruitment and development, through bureaucratic performance management systems, to the more ambitious initiatives to manipulate employee attitudes and commitment through culture change and 'empowerment'.

iii. **Deregulation:** Closely allied with the use of new techniques of labour organization and control has been a renewed emphasis on the management prerogative and the need for managers to respond to market signals without hindrance.

iv. **Rhetoric and deceit**: This has to do with a seeming decline in management integrity and what many characterize as the manipulative tendencies of HRM.

These categories have wide-ranging implications for individual employees and for the HR function and reflects the increasing importance of HRM to organizational survival and the increasing role multiplicity of HR practitioners, engendered by the continued modernization and redefinition of the HR function. In this regard, works by Archer (1986), Edwards and Bennett (1987) and Wallace (1985), all illustrate a common theme. The core theme is that HRM practitioners engage in a high degree of role multiplicity, which influences the opportunity for ethical dilemmas to occur.

Adapting Wooten's (2001: 162) position, role multiplicity here refers to the enactment of many different functions, duties, and responsibilities on the part of HRM professionals. For instance, HRM professionals set up and monitor policies, make decisions impinging upon employee rights, enforce employment law, educate organizational participants, and contract with external agencies for services, and so on. Additionally, they are frequently seen by others as an organizational watchdog for all other organizational participants (Edwards & Bennett, 1987; cited in Wooten, 2001). Thus, the role of the HRM professional and their subsequent boundary spanning functions provide a rich context in which ethical dilemmas reside.

Several authors have articulated the broad-range of ethical challenges and dilemmas in contemporary HRM. For instance, Gandz and Hayes (1988) noted issues such as discrimination, psychological testing, antiunion activity, work design, employment security, privacy, negotiation tactics, disciplinary practices, worker displacement, and confidentiality to be among the most difficult of ethical problems in HRM practice.

McLagan (1989) has noted many of these same issues in the HRM subfield of HR development, inclusive of management of diversity differences, balancing of organizational and individual needs, respect for copyrights/intellectual property issues, and ensuring truth in data and recommendations, to name a few. Similarly, Ekuma and Smith (2012) and Wooten (2001) in their different works have also identified five applicable ethical dilemma categories to include:

i. Misrepresentation and collusion.

ii. Misuse of data.

iii. Manipulation and coercion.

iv. Value and goal conflict, and

v. Technical ineptness.

An interesting point to note from these categorizations is that there is a broad agreement amongst scholars (e.g. Ekuma & Smith, 2012; Wooten,

2001), that these dilemmas arise primarily because of the dual nature of HRM in contemporary organizations.

6. DEVELOPING A RESEARCH AGENDA

While it has to be acknowledged that the ethical concern in the employment relationship, particularly with regards to distributive and organizational justice and more recently on employee rights, has been an enduring academic and professional interest for decades, right from the days of industrial relations and has continued till the present times of modern HRM. The tremendous changes that have taken place in HRM over the past few years, means that most of these issues have changed profoundly and now require fresh insights; even as new challenges continue to emerge. Winstanley and Woodall (2000: 8) aptly capture the trend thus:

Although many aspects of the traditional ethical agenda in HRM are still relevant, evolving approaches to HRM suggest that it is more than 'new wine in old bottles' and that with the evolving new approaches, come new ethical problems. What makes this analysis even more complex is that there is not just one new model of HRM but many.

Indeed, a number of themes, including: the preoccupation with flexibility, commitment, culture, performance and work-life balance, all appear to be associated with contemporary HRM. These themes raise a number of 'unique' ethical issues which have continued to bewilder practitioners and scholars alike. Winstanley and Woodall (2000: 8) highlight some of these peculair ethical issues to include:

- Flexibility in variable pay systems or in the contract of employment and 'high commitment' work practices raises ethical questions about practices as varied as 'presenteeism' and long working hours.
- Performance management systems based on stretch' targets and close surveillance and control place increasing emphasis on processes for evaluating, grading and classifying individuals.
- A desire to 'capture hearts and minds' in the service of corporate goals has extended the focus of training and development activity beyond the mere acquisition of knowledge and skills into shaping values and attitudes, by means of new techniques of value and culture change.

The above issues are not entirely new to management scholars, but changing behaviors and the nature of contemporary business environment arguably introduce additional dimensions beyond a concern with justice, stake-holding, rights or even universalism. Fresh questions are also going to continue to be raised about the scope of employer duty of care, about individual rights to autonomy, privacy, dignity and self-esteem, and the boundaries between organizational demands and employee subjectivity (Winstaley, et al., 1996). In order to illuminate these issues, it is suggested here that scholars should focus on a number of critical, but related research agendas as follows:

1. Critically explore the contemporary HR function with a view of highlighting major ethical issues and challenges confronting practitioners.
2. Investigate and illustrate the precise nature and current dilemmas faced by HR practitioners in the workplace.
3. Clearly articulate and recommend a practicable strategies and a frameworks for dealing with ethical dilemmas in the course of HR practice.

It is now the responsibility of scholars and practitioners, particularly those in HRM, to articulate the best possible approaches to address these emerging and contemporary ethical agenda. In this regard and consistent with Winstanley and Woodall's (2000: 8) position, it could be argued that there are two key issues involved:

1. Concerns about the nature of ethical inquiry and its relation to action.
2. Concerns about the ethical frameworks to be employed.

In addition to the above concerns, I suggest that issues about professionalism and adherence to basic code of conducts should also be high on the agenda, not only as a theoretical model for understanding the basic acceptable standards of practice and for decision making, but also as practical guide to curb excesses and deal effectively with unqualified persons and/or substandard/unethical behaviour.

So, as a starting point, writers on human resource management and practitioners alike could benefit from acquainting themselves with relevant business ethics frameworks. As this review indicates, there are many ethical frameworks in relation to human resource management and there is no one best approaches. An awareness of these frameworks is the basic level for engagement in a productive debate on how best to tackle the emerging exigencies of contemporary ethical issues in HRM (Winstaley, et al., 1996).

The second level of engagement with HRM ethics, is for human resource writers and practitioners to utilize appropriate ethical frameworks to help explain and analyse the nature of the changes taking place in the employment relationship, and its ethical dimensions (Winstaley, Woodhall & Heery, 1996). The third level is to use the frameworks more prescriptively in action – at least as a guide, for practitioners to use them as a basis for action. At the very least we could identify the 'process of reasoning' by which HR policy decisions and acts are justified, by focusing on two broad categories: dilemmas arising from HR functional areas; and dilemmas arising from organisational policies, culture and strategy.

7. CONCLUSIONS

It is evident from what has been said in this paper, that the employment of people in contemporary times gives rise to unique and critical ethical dilemmas on the part of HR practitioners, which have not been sufficiently addressed in the extant literature. The ever increasing competitive pressures and greater employment scrutiny, means that human resource professionals must make and execute decisions with several ethical implications, which may result to serious dilemmas for the practitioners. Indeed, beyond the traditional issues of confidentiality, fair treatment and honesty, HR must also deal with other controversial, but equally important issues, including affirmative action, health and safety, harassment as well as functional related issues including redundancies, downsizing, pay and benefits, promotion and so on.

In all of this, the critical need to positively engage employees for increased productivity must not be undermined in the process. It is therefore, the responsibility of the HR function to strike this delicate balance of ensuring employees' welfare, while not compromising the need for increased organisational efficiency and improved productivity. To achieve this, the practitioner must endure profound pressures, make very difficult decisions and encounter very many ethical challenges which sometimes results in dilemmas. There is the need for a better understanding of the nature and dynamics of these dilemmas, in order to identify frameworks and suggest strategies that will at least help HR professionals in dealing with ethical dilemmas in HR practice. This paper identified a number of broad research agendas that scholars and practitioners should now focus to help towards gaining this understanding and in the identification process. While by no means exhaustive nor prescriptive, the intention is that the agendas identified here, will engender more debates and lead to the exploration of practitioners' dilemmas from a multidimensional perspective. This will allow researchers and other interested parties to better understand the dynamics of HR ethics and its associated dilemmas, so as to afford HR professionals with strategies to consider when resolving ethical challenges or dilemmas in the workplace.

REFERENCES

1. Abreu, J. L., and Badii, M. H. (2006). Proposal of an ethical model for human resource management (Propuesta de un modelo ético de gestión de recursos humanos). *International Journal of Good Conscience.* 1(1) March: 102a – 119a.

2. Appelbaum, S. H., Deguire, K. J., & Lay. M. (2005). The relationship of ethical climate to deviant work behaviour. *Corporate Governance,* 5(4), 43 – 55.

3. Archer, E. R. (1986). Human resource professionalism: an unexpected source of conflict. *Personnel Administrator,* 31, 97-104.

4. Beauchamp, T. L., and Bowie, N. E. (1983). *Ethical theory and business,* 2nd ed. New Jersey: Prentice Hall.

5. Billington, R. (2003). *Living philosophy: an introduction to moral thought,* 3rd ed. London: Routledge.

6. Birsch, D. (1995). The universal drug testing of employees. *Business and Professional Ethics Journal, 14,* 43-60.

7. Carse, A. (1996). Facing up to moral perils: the virtues of care in bioethics, in S. Gordon, P. Benner and N. Noddings (eds). *Care-giving: readings in knowledge, practice, ethics and politics.* Philadelphia: University of Pennsylvania Press.

8. Carson, A. S. (1995). Drug testing and privacy: why contract arguments do not work. *Business and Professional Ethics Journal, 14,* 3-22.

9. http://dx.doi.org/10.5840/bpej19951449.

10. Chartered Institute Personnel and Development (CIPD). (2010). Employee relations: an overview. *Fact-Sheet* (July). London: CIPD.

11. Christy, C., & Brown, E. (2009). Ethics and diversity in human resource management, in C. Rayner and Derek Adam – Smith, *Leading and managing people,* 2nd ed. London: CIPD, 66 – 85.

12. Cornelius, N & Gagnon, S. (1999). From ethics 'by proxy' to ethics in action: new approaches to understanding HR and ethics. *Business Ethics: A European Review,* 8(4)(Oct.): 225 – 235.

13. http://dx.doi.org/10.1111/1467-8608.00157.

14. Cornock, C., and Johns, T. (1995). *Ethical leadership.* London: CIPD.

15. Dachler, H. P., & Enderle, G. (1989). Epistemological and ethical considerations in conceptualizing and implementing human resource management. *Journal of Business Ethics, 8,* 597-606.

16. http://dx.doi.org/10.1007/BF00383028.

17. Danley, J., Harrick, E., Schaefer, D., Strickland, D., Sullivan, G. (1996). HR's view of ethics in the work place: are the barbarians at the gate? *Journal of Business Ethics*, 15(3) (Mar.): 273-285. (Online). Available at:<http://www.jstor.org/stable/25072753> [Accessed 14 March 2011].

18. Deshpande, S. P. (1996). Ethical climate and the link between success and ethical behavior: an empirical investigation of a non-profit organization. *Journal of Business Ethics,* 15(3) (March): 315-320. (Online). Available at:<http://www.jstor.org/stable/25072757> [Accessed 14 March 2011].

19. Edwards, G., & Bennett, K. (1987). Ethics and HR: standards in practice. *Personnel Administrator, 32,* 62-66.

20. Ekuma, K. J & Smith, J. (2012). *Ethics in human resource management: Contemporary dilemmas of the practitioner.* Saarbrücken, Germany: LAMBERT Academic Publishing.

21. Foote, D. (2001). The question of ethical hypocrisy in human resource management in the U.K. and Irish charity sectors. *Journal of Business Ethics*, 34(1) (Nov.): 25-38.

22. Fredman, M. (1970). The social responsibility of business is to increase its profits. *Times Magazine.* New York, 13 September, pp 32ff.

23. Gandz, J., & Hayes, N. (1988). Teaching business ethics. *Journal of Business Ethics, 19,* 657-669.

24. http://dx.doi.org/10.1007/BF00382975

25. Greenwood, M.R. (2002). Ethics and HRM: a review and conceptual analysis. *Journal of Business Ethics,* 36(3) (March): 261-278.

26. Greenwood, M.R., 2002. Ethics and HRM: a review and conceptual analysis. *Journal of Business Ethics,* 36(3) (March): 261-278.

27. Guest, D. (1997). Human resource management and performance: a review and research agenda. *The International Journal of Human Resource Management*, 8(3): 263-276.

28. http://dx.doi.org/10.1080/095851997341630

29. Henderson, M. J. (1997). Ethical outsourcing in U.K. financial services: employee rights. *Business Ethics,* 6, 110-118.

30. http://dx.doi.org/10.1111/1467-8608.00058

31. Huselid, M. A. (1995). The impact of human resource management practices on turnover, productivity, and corporate financial performance. *Academy of Management Journal,* 38(3): 635-672.

32. http://dx.doi.org/10.2307/256741

33. Johnson, G., and Scholes, K. (1997). *Exploring corporate strategy,* 4th ed. Hemel Hempstead: Prentice Hall.

34. Kew, J., and Stradwick, J. (2008). *Business environment: managing in a strategic context.* London: CIPD.

35. Kupfer, J. (1993). The ethics of genetic screening in the workplace. *Business Ethics Quarterly,* 3, 17-25.

36. Legge, K. (1995). *Human resource management: rhetorics and realities.* Basingstoke: Macmillan.

37. Legge, K. (1998). The morality of HRM, in C, Mabey, D, Skinner & T, Clark (eds.), *Experiencing human resource management.* London: Sage.

38. Mathis, R. L., & Jackson, J. H. (1997). *Human resource management,* 8th ed. St. Paul: West Publishing.

39. McLagan, P. (1989). Models for HRD practice. *Training and Development Journal, 43,* 49-59.

40. Palmer, G. (2007). Socio-political theory and ethics in HRM, *in* A Pinnington., R Macklin, & T Campbell, (eds.) *Human resource management: ethics and employment.* New York: Oxford University Press.

41. Payne, S. L. (1994). Epistemological and ethical development for human resource professionals. *Business and Professional Ethics Journal,* 13, 33-56.http://dx.doi.org/10.5840/bpej199413316.

42. Pinnington, A., Macklin, R., & Campbell, T. (eds.). (2007). *Human resource management: ethics and employment.* New York: Oxford University Press.

43. Purcell, J. (1999). Best practice and best fit: chimera or cul-de-sac?' *Human Resource Management Journal* 9(3): 26-41.

44. http://dx.doi.org/10.1111/j.1748-8583.1999.tb00201.x

45. Raibom, C., & Payne, D. (1996). TQM: just what the ethicist ordered. *Journal of Business Ethics,* 15: 563-572.

46. Senneth, R. (1998). *The corrosion of character: the personal consequences of work in the new capitalism.* London: Norton.

47. Stewart, J., and Rigg, C. (2011). *Learning and talent development.* London: CIPD.

48. Taskin, L., & Devos, V. (2005). Paradoxes from the individualization of human resource management: the case of telework. *Journal of Business Ethics*, 62(1) (Nov.): 13-24. (Online). Available at: http://www.jstor.org/stable/25123639 [Accessed 14 March 2011].

49. Torrington, D., and Hall, L. (1998). *Human resource management*, 4th ed. Hemel Hempstead: Prentice-Hall Europe.

50. Wiley, C. (1998). Re-examining perceived ethics issues and ethics roles among employment managers. *Journal of Business Ethics*, 17(2) (Jan): 147 – 161.http://dx.doi.org/10.1023/A:1005795731002.

51. Wiley, C. (2000). Ethical standards for human resource management professionals: a comparative analysis of five major codes. *Journal of Business Ethics*, 25: 93 – 114.

52. Winstanley, D. and Woodall, J. (2000b). (eds). *Ethical issues in contemporary human resource management.* Basingstoke: Macmillan.

53. Winstanley, D., & Woodall, J. (2000). The ethical dimension of human resource management. *Human Resource Management Journal*, 10(2): 5–20.

54. Winstanley, D., Woodall, J., and Heery, E. (1996). The agenda for ethics in HRM. *Business Ethics*, 5(4): 187 - 194.

55. http://dx.doi.org/10.1111/j.1467-8608.1996.tb00158.x

56. Wooten, K.C. (2001). Ethical dilemmas in human resource management: an application of a multidimensional framework, a unifying taxonomy, and applicable codes. *Human Resource Management Review*, 11, 159 – 175.

57. http://dx.doi.org/10.1016/S1053-4822(00)00045-0.

Chapter 2

Strategic Human Resource Management in U.S. Luxury Resorts—A Case Study

Marcia Taylor & Dori Finley

Department of Hospitality Management, East Carolina University, Greenville, NC

ABSTRACT

A labor shortage has been experienced in the hospitality industry and is predicted to continue into the future with a greater impact on luxury resorts. Resort managers typically look to human resource (HR) directors to develop strategies to solve this problem. The co-alignment model can give managers a competitive advantage in the marketplace. This study presents the results of a case study of five luxury resorts in North Carolina. HR directors identified forces driving change in the environment, strategy choices, firm structure, and outcomes reflected in firm performance. There was little evidence that co-alignment was being used as a basis for planning.

Keywords : Strategic management, co-alignment, human resources, resorts

INTRODUCTION

The labor shortage has been recognized nationally as a major force driving change for decades and is predicted to continue into the future with the shortage having greater impact on the hospitality industry (Terry, 2005). This shortage is even more amplified for resorts that are typically located in remote areas with a high cost of living, low unemployment rates, and a seasonal need for employees (Angelo & Vladimir, 2004). These three factors have led to increased turnover and higher overall costs for resorts. In the past 3 years, there have been many media reports on the seriousness of the labor shortage in seasonal resorts due to changes in immigration laws (Berta, 2004; Hedlund, 2004).

Solving the labor shortage problem is the responsibility of the human resources (HR) department. Traditionally, this department has served as support for operations and was viewed as a funnel to provide workers. Administrative functions of the department were viewed as the only contribution of HR to the total organization. This view changed due to a movement in business and industry that treated HR as human capital. Today, in some companies, the HR department has been viewed as a source of competitive advantage and has become a strategic partner at the executive level (Kearns, 2004). This has served to differentiate companies with a strategic HR emphasis from those without. According to Cooper (2005), what differentiates great companies from their peers is the ability to hire, develop, and retain the best people. The effect of a strategic emphasis in hospitality HR departments has not been documented in the United States, thus the need for research in this area.

This article reports the results of a case study of five luxury resorts in North Carolina. The issues investigated in this article include:

1. recognition by luxury resort HR managers of the forces driving change in the environment;
2. competitive methods being utilized to solve the labor shortage;
3. resources allocated to these competitive methods;
4. alignment of the three elements of the model with firm performance.

REVIEW OF THE LITERATURE

Beginning with a review of literature on the labor shortage issue, changes in HR, and the co-alignment model, this study reports results using the elements of the co-alignment model, and identifies solutions and strategies used by HR directors in luxury resorts. The hotel industry has been segmented by Smith Travel Research. The top three categories are luxury, upper upscale, and upscale (O'Neill, 2006). However, luxury has been redefined as an experience versus a product (Schiller, 2006). This definition was repeated by Weinstein (2002) who defined luxury as employees who deliver service that anticipates needs and who pay attention to guests.

Labor Shortage in Hospitality Management

An evaporating pool of workers, coupled with other labor concerns, may be the biggest challenge that the hospitality industry faces. The labor force is expected to grow as little as 16% over the next 20 years, compared with the 50% growth of the previous two decades (Gillette, 1996). The Employment Policy Foundation forecasts a deficit of labor needed versus labor available of 10 million workers in 2015 growing to 35 million by 2030 (Holt, 2006). The National Restaurant Association predicts that the industry will need to fill 1.4 million new jobs by 2010—more jobs than people willing to take them (Berta, 2002). The International Society of Hospitality Consultants has identified a shortage of labor and skills as the number-one issue for 2007 (ISHC, 2006).

Traditionally, seasonal resorts in the United States depended heavily on workers from other countries for restaurants, housekeeping, and landscaping positions (Taylor, Finley, & Calvert, 2005). Even when there were plenty of potential employees available, turnover has been a problem in the industry, hovering at 100% for line workers. The labor shortage in the hospitality industry has been recognized as a major force driving change for decades; however, the industry has failed to identify solutions to address this issue. Solutions that could be implemented to ease the shortage of labor in the industry include outsourcing, improving productivity, recruiting in target markets, developing attractive employment policies, marketing as employer of choice, and increasing skills of employees (Holt, 2006).

According to Ettedgui (2006), the best luxury hotels are known for providing exceptional services and for the sincerity of the people who deliver those services. This is true in luxury resorts where the delivery of quality service is usually the number-one priority. In a model of impediments to improvements in service quality in luxury hotels, Presbury, Fitzgerald, andChapman (2005) identified four broad categories of impediments: (1) budget constraints, (2) staff attitude, (3) lack of mentoring, and (4) high customer expectations. Two of these categories, staff attitude and lack of mentoring, are impacted by HR policies and procedures. The authors indicate that a lack of leadership, inexperienced managers, high turnover, recruitment procedures, and a lack of service ethic in the organization are issues that contribute to the lack of mentoring and staff attitude. All of these issues impact labor turnover and should be addressed by hospitality HR directors.

Strategic Management in HR

Historically, the role of HR in a hospitality management company has been administrative in nature. Fulford and Enz (1995)documented this administrative definition, which they called personnel administration, in the HR department of a multi-unit restaurant chain. A national trend in HR is to move from the administrative role to the incorporation of HR in strategic planning. This movement was aided by the development of the concept of human capital or human assets in an organization. Human capital is defined as including "skills, judgment, and intelligence of the firm's employees" (Barney & Wright, 1998, p. 32). In a Norwegian hotel chain, Engstrom, Westnes, and Westnes (2003) identified human capital as one of three components of intellectual capital; the other components were customer capital and structural capital. These authors identified measures for human capital as competence, improvement systems, intellectual agility, performance, and attitude and motivation.

One study that investigated human capital in U.S. hotels identified a Hospitality Human Capital Process Model (Young, McManus, & Canale, 2005). The three components of the model are (1) service-oriented employees, (2) empowered employees, and (3) committed employees. Developing service-oriented employees requires training on guest expectations, an appropriate hiring process, and a service-oriented culture. Developing empowered employees requires training on problem solving; shared values, norms, and goals; in addition to an appropriate hiring process and service oriented culture. Managerial activities that develop committed employees include nurturing psychological bonds, treating employees fairly, meeting employee expectations, in addition to a hiring process that selects "best fit" employees. This model delineates a comprehensive program for maximizing hospitality HR.

Building on the view of HR as human capital, strategic human resources management (SHRM) includes approaches for matching people to business strategies (Miles & Snow, 1984). A model of development of the HR function in organizations includes five levels divided into two categories according to Kearns (2004). The two categories are where employees are seen as a cost/resource and where employees are seen as a competitive advantage. When employees are seen as a competitive advantage, HR becomes a strategic partner responsible for getting the maximum value from the company employees. Strategic HR managers see the workforce as a source of strategic advantage, not a cost to minimize (Pfeffer, 2005).

An increase in the strategic approach to human resources management in the U.K. hotel industry was reported byHoque (1999) and was recommended for multi-unit restaurants by Fulford and Enz (1995). It was proposed by Guest (1987)that the integration of HR into strategic plans supported by policies, a culture that stresses the importance of HR, and employee commitment will lead to the successful implementation of those strategic plans.

The results of a project to investigate the impact of technical (the administrative role) HR and SHRM on firm performance indicated that SHRM has more positive impact on firm performance than technical HR (Huselid, Jackson, & Schuler, 1997). These researchers also linked SHRM to competitive advantage and, in turn, to business performance.Olsen, West, and Tse (2007) use the co-alignment model to demonstrate the significance of this link.

Co-Alignment Model

As early as 1983, strategic management has been recognized as important for managers in the hospitality industry (Reichel, 1983). More recently Kim and Oh (2004) recommended the use of a comprehensive, integrated strategic-management method to give hotels a competitive advantage. The co-alignment principle is a strategic-management theory that implies that companies can gain a sustainable competitive advantage if management adopts the principles of the theory in their everyday operations (Olsen et al., 2007). In support of this theory, Olsen et al. (2007) developed a model that consists of four constructs necessary in achieving co-alignment within business. According to the co-alignment principle model, if management can (1) identify opportunities that exist in the forces driving change in the environment, (2) invest in competitive methods (strategy choices) that take advantage of these opportunities, and (3) allocate resources to those methods that create the greatest value, then they can (4) achieve sustainable competitive advantage (Olsen et al., 2007; see Figure 1).

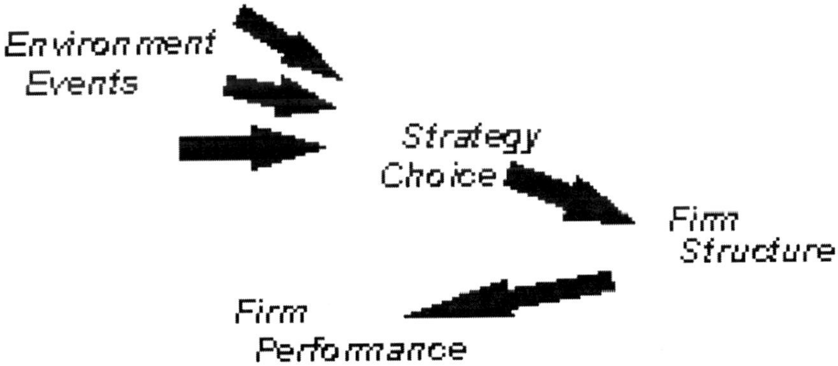

Figure 1. Co-alignment Model Adapted from Olsen et al. (2007)

A firm's success depends on the management's awareness of the environment and ability to identify and adapt to changes. According to Olsen et al. (2007), the co-alignment model is similar to other strategic management concepts, but it is future oriented. The application of the co-alignment model to the hospitality industry has been reported by Olsen et al. (2007) and in many unpublished dissertations. Typically, the hotel industry is considered a "copycat industry." Hoteliers are quick to copy innovations by others. Adopting the co-alignment model can give a resort manager a competitive advantage in the marketplace. Due to the competition for luxury travelers and the short lifespan of luxury services, application of this model is appropriate for managers in luxury resorts.

PURPOSE OF STUDY

The purpose of this study was to answer two research questions: (1) Are luxury resorts investing in competitive methods to take advantage of the opportunities that exist in the forces driving change in the environment? (2) Are luxury resorts allocating resources to those competitive methods that create the greatest value? Specifically, the study sought to find out if luxury resort HR managers recognize the forces driving change in the environment, the competitive methods they are utilizing to solve the labor shortage, how they are allocating resources to these competitive methods, and the performance indicators as a result of implementation.

METHODOLOGY

To answer the research questions, the case study method with multiple cases as described by Yin (1993) was used. The case study method has been demonstrated as appropriate in testing the co-alignment model because researchers must enter into the domain of the firm and study it in depth in order to understand the complexities of the situation (Taylor & Olsen, 2006). The use of face-to-face interviews has proved effective in testing the co-alignment model because the subjects may not be familiar with the concepts included in the co-alignment model, and interviewing allows researchers to probe and use questions to get a valid response. Other researchers (mostly unpublished dissertations) have used the case study method in investigating the co-alignment model. In addition, Aung (2000) used the case study method to identify the core competencies of the Accor hotel chain.

The focus for this study was 4-star resorts in North Carolina. Seven resorts were identified in the *Official 2005 North Carolina Travel Guide* (2005). After contacting the HR directors at these resorts, five of the seven HR directors agreed to participate in the study. Yin (1993) suggested the use of multiple cases be viewed as multiple experiments and not multiple respondents to a survey. The consensus for numbers of cases falls between two and four as the minimum and ten and fifteen as the maximum (Perry, 1998). Therefore, the five cases were considered to be adequate for this case study.

A structured questionnaire consisting of twenty-seven open-ended questions was developed to serve as the basis of the face-to-face interviews with each director on location at the resort. Six of the questions were descriptive of the resort. Five questions sought to identify the forces in the environment identified by the HR director as driving change in the hotel industry. These forces also have the potential of affecting the resort in the future and are contributing to problems in HR at the resort. Four questions addressed the strategy choices or competitive methods utilized by the resorts. Specifically, the HR directors were asked to identify the resort's competitive methods and what was included in each method. They were also asked how competitive methods were chosen and which methods were perceived as adding the most value to the resort. The next five questions sought to identify how capital and HR were allocated to the competitive methods. Firm performance in terms of financial performance was identified in the final eight questions.

The questionnaire was e-mailed to the HR director for preparation prior to the face-to-face interview. The interviews ranged from 1.5 to 2.5 hours in length. In addition to the interview, the researchers were able to observe the implemented strategies at each property to confirm information received during the interviews. Interviews were transcribed and content analysis was used to evaluate the responses (Neuendorf, 2002).

RESULTS AND DISCUSSION

The results of the case study include demographic information about the resorts, content analysis for components of the co-alignment model, and evaluation of the alignment/nonalignment of the components.

Characteristics of the Sample

There was a wide cross-section of locations for the resorts ranging from mountain to ocean and from city to countryside. All had convention and meeting spaces, all but one had spas, and all but one had golf courses. Three resorts were medium-sized hotels with under 300 rooms. One resort had more than 500 rooms. The resorts were not chain owned or operated. Two of the resorts would be considered historical properties; however, all of the resorts are recognized as luxury resorts in North Carolina. Only one of the resorts operated as a seasonal resort.

Forces in the Environment Driving Change

The first component of the co-alignment model is the identification of forces in the environment that are driving change. All HR directors identified similar environmental forces driving change in their resorts (see Table 1). All directors cited economic issues related to labor and guests as forces driving change in the environment. The labor issues included immigration and diversity of the workforce, unemployment rates, energy costs (gasoline) for employees, changes in benefits and their increasing costs, and the generational mix. Economic issues related to guests included growth in family travel and more choices of resorts as a result of new competition. Other issues that were identified included the impact of technology and environmental concerns.

Table 1. Content Analysis Results for Forces in the Environment, Strategy Choices/Competitive Methods, and Firm Structure

Forces in the Environment	Strategy Choices/ Competitive Methods	Firm Structure
Economic Issues—Labor		
Unemployment rates	H2B, J-1 Visas	Increased HR budgets
	Competitive benefits and salaries	Changing management structure to compliment capital investment
	Leadership development and increased training	HR as strategic partner
	Developing a company culture	Culture committee
	Building loyalty among employees	
	Incentive programs	
	Redefining full-time employment	
	Extending seasons	
Benefits increasing in cost and changing in nature	Redefining full-time employment	
	Flextime	
	Child care	
	Job sharing	
Immigration and diversity of the work force		
Energy costs for employees		
Generational mix		
Economic Issues—Guests		
Growth in family travel due to increasing energy costs or value of time	Renovated facilities	Focused marketing
	Added amenities	
	Increased service quality	
	Package pricing	
New competition	Renovated facilities	Focused marketing
	Added amenities	
	Increased service quality	
	Package pricing	
	Leadership development and increased training	
	Cobranding	
Other Issues		
Technology		
Environmental		

Application of the co-alignment model to the five cases indicated that the HR directors were monitoring forces in the environment. However, there was a range of focus used by the directors. All but one of the HR directors was focusing on the environment at the national and international level, which contrasted with the other director who was focused locally. As suggested by Schuler and Jackson (2005), looking more broadly will assist HR directors in becoming strategic partners in operations decisions. Three of the HR directors had participated in an executive-level strategic-planning process that identified forces in the environment that were impacting their resorts. While at the other end of the continuum, the HR directors monitored local events.

Strategy Choices/Competitive Methods

The second component of the co-alignment model is the utilization of strategy choices/competitive methods. The HR directors identified their strategy choices/competitive methods for two of the labor issues and both of the guests' issues. No strategy choices/competitive methods were identified for the technology and environmental issues (see Table 1). Strategies selected to address the unemployment rate issue included use of H2B and H-1 visa employees (guest workers or temporary workers), offering competitive benefits and salaries, leadership development and increased training, developing a company culture and loyalty among employees, offering incentive programs, redefining full-time employment, and extending the seasons with off-season offerings. Strategies utilized to address the increased cost and changing nature of benefits included redefining full-time employment, flextime, child care, and job sharing. No strategies were implemented to address the HR-related issues of immigration and diversity of the workforce, energy costs for employees, or the generational mix.

The growth in family travel was addressed by renovated facilities, added amenities, increased service quality, and package pricing. New competition resulted in renovated facilities, added amenities, increased service quality, package pricing, leadership development, and increased training and co-branding. The strategies identified for the two changes in guest needs were the same with new competition generating the need for leadership development, increased training, and co-branding.

It can be concluded from these findings that HR directors are making strategic choices or developing competitive methods to address the forces driving change in the environment. Six methods were developed to address more than one force in the environment, which is similar to the results in other studies (Young et al., 2005). The range in strategy choices varied from comprehensive to piecemeal. Three of the HR directors described "People

Strategies," which were very comprehensive programs, while others listed separate competitive methods with no linkage. The range of the methods was from those based on data analysis, as in the total rewards strategy described by Fischer, Gross, and Friedman (2003), to others based on "best guess." For example, in one resort, for every dollar invested in their employees an increase of three dollars in profit was generated. At another resort, any increase in revenue was interpreted as meaning that they were doing the right thing.

Competitive strategies that impacted employees directly included redefining full-time work, offering competitive wages and benefits, creating loyalty in employees, and a variety of changes to encourage a more-stable workforce. In one resort, the minimum number of hours required for an employee to receive full-time benefits was decreased from 40 hours per week to 30 hours per week. This change allowed employees to stay on the payroll and receive benefits during slower seasons. A variety of wellness programs and incentive programs were described as methods for increasing the health and well-being of employees and to reward employees for contributing to the bottom line of the resort.

The HR director at one resort reported treating employees like family as a method used to create loyalty in the workforce. Examples of how a family environment type of work culture was created included fundraisers, company support for employees with need, supporting the community with funds, and paid time for employees to assist community organizations. Changes made to encourage a more stable workforce included extending the seasons by offering themed weekend events, promoting job sharing, and offering flextime. It is interesting to note that none of the resorts used outsourcing to solve labor problems and the implementation of all of these changes decreased the need to use guest workers.

Competitive HR strategies that were implemented in response to the changes in guest needs included more training and leadership development, and developing a full-time, year-round workforce. The increase in training allowed the resorts to capitalize on the strengths of their current employees. By decreasing the need for part-time or seasonal workers, the resorts could offer consistent quality service as demanded by guests at a luxury resort.

Firm Structure

The third component of the co-alignment model is the firm structure required to implement the strategy choices/competitive methods that have been selected. Results in Table 1 indicate increased HR budgets, change of management structure to compliment capital investment, involving HR as a strategic partner, and establishing a culture committee as needed to address the unemployment rate issue. Focused marketing was implemented to address both of the guest issues.

Structural changes that were identified by the HR directors were not as comprehensive as the strategy choices. In addition, they were not tied directly to a strategy choice as the co-alignment model would indicate. The comprehensiveness of the responses received from the HR directors ranged from all of the structural changes included in Table 1 to listing only one—increased HR budgets.

Firm Performance

The final component of the co-alignment model is the firm performance measures that are used to determine the impact of the changes. Traditional measures of performance in the lodging industry (average daily rate [ADR], revenue per available Room [RevPar], turnover, etc.) were identified by all but two of the HR directors. These results are similar to those reported byMandelbaum (2006). It was interesting to note that almost all of the respondents did not have access to the operating data on a regular basis and in no case could alignment be determined. The annual ADR ranged from $185 to $300, annual occupancy rates ranged from 59% to 96%, and annual RevPar ranged from $125 to $559. Only one HR director reported seasonal data. Employee data included payroll percentages ranging from 32% to 38.5%, employees to rooms ranging from 1.1 to 1.9, and turnover percentages ranging from 16.5% to 65%.

The wide range in RevPar, occupancy rates and turnover were due to data from one seasonal resort. Turnover rates were also impacted by the use of temporary guest workers. The HR directors reported that turnover rates were not increasing. This trend is different from a study by the Society for Human Resource Management where 38% of the members reported increasing turnover rates (Feeney, 2007). While it is difficult to link these data as outcomes for a strategy choice and change in firm structure, there was a trend for lower turnover rates in the resorts with a more comprehensive "People Strategy."

IMPLICATIONS AND CONCLUSION

Succeeding in the resort industry requires alignment between the four elements of the co-alignment model: (1) the environment, (2) strategy choice(s), and (3) firm structure, which leads to an outcome reflected in the (4) firm performance (Olsen et al., 2007). From the interviews with the HR directors, it was clear that the forces driving change in the hospitality industry environment, and more specifically the resort industry, were identified as economic issues related to labor and guests. Each of the HR directors identified forces in the environment; however, the critical difference was in looking locally versus nationally for changes. In two of the five resorts, the HR director was viewed as a source for strategies to address the changes in the environment. In the other three resorts, the HR director was not. In all but one of the resorts creative structural changes are being made to achieve competitive advantage.

When applying the co-alignment model in this case study, there is very little evidence that co-alignment is being used as a basis for planning. The forces driving change in the environment and competitive methods were identified by all HR directors. The allocation of resources to these methods was not clearly delineated. In addition, the outcome measures and the effectiveness of each method were not identified by the directors. Use of the co-alignment model could give these resorts a competitive advantage due to their constant need for outstanding service as expected in a luxury resort.

Although limited to resorts in only one southeastern U.S. state, the findings indicate the need to duplicate this research project with data gathered from more resorts across the country. Through additional case studies, the use of SHRM and the co-alignment model could be documented as providing a competitive advantage in resorts. Further research needs to be conducted using the general manager as the source for information. The results of such a study would be strengthened by their knowledge of performance measures.

Despite the limited number of cases in this study, the information gathered should help HR directors in the hotel industry as they identify and react to the forces driving change in the environment.

REFERENCES

1. Angelo, R. and Vladimir, A. 2004. *Hospitality today: An introduction*, Lansing, MI: Educational Institute of the American Hotel and Lodging Association.

2. Aung, M. 2000. The Accor multinational hotel chain in an emerging market: Through the lens of the core competency concept. *Service Industries Journal*, 20(3): 43–60.

3. Barney, J. B. and Wright, P. M. 1998. On becoming a strategic partner: The role of human resources in gaining competitive advantage. *Human Resource Management*, 37(1): 31–46.

4. Berta, D. 2002. Special report—opportunity through tears: Immigration. *Nation's Restaurant News, 36*, 9: 122–123.

5. Berta, D. 2004. U.S. to foreign workers: Stay home for the summer. *Nation's Restaurant News*, 38: 1–3. 15

6. Cooper, K. 2005. The missing link in people strategies. *Customer Inter@ction Solutions, 24*, 5: 48–50.

7. Engstrom, T. E. J., Westnes, P. and Westnes, S. F. 2003. Evaluating intellectual capital in the hotel industry. *Journal of Intellectual Capital*, 4(3): 287–303.

8. Ettedgui, E. 2006. Striving to offer quality in a flawless manner. *Financial Times*, September 18: 5

9. Feeney, P. 2007. *Hotel recruitment: Is a talent shortage looming?* Retrieved January 4, 2007 from,http://www.hotelexecutive.com/bus_rev/pub/002/744.asp*Hotelexecutive.com*

10. Fisher, K., Gross, S. E. and Friedman, H. M. 2003. Marriott makes the business case for an innovative total rewards strategy. *Journal of Organizational Excellence*, 22(2): 19

11. Fulford, M. D. and Enz, C. A. 1995. Human resources as a strategic partner in multiunit-restaurants. *Cornell Hotel and Restaurant Administration Quarterly*, 36(3): 24–29.

12. Gillette, B. 1996. Labor pains (labor shortage in the hospitality industry). *Hotel and Motel Management*, 211(4): 24–26.

13. Guest, D. E. 1987. Human resource management and industrial relations. *Journal of Management Studies*, 24(5): 503–521.

14. Hedlund, S. 2004. Restaurateurs and processors brace for a potential labor shortage: Employers urge lawmakers to lift or raise the annual cap on foreign seasonal workers. *Seafood Business*, 23(5): 34

15. Holt, K. 2006. Workforce plan critical to survival during labor deficit. *Hotel and Motel Management*, 221(16): 27

16. Hoque, K. 1999. New approaches to HRM in the UK hotel industry. *Human Resource Management Journal*, 9(2): 64–76.

17. Huselid, M. S., Jackson, S. E. and Schuler, R. S. 1997. Technical and strategic human resource management effectiveness as determinants of firm performance. *Academy of Management Journal*, 40(1): 171–188.

18. ISHC. 2006. Top ten issues in the hospitality industry for 2007. *Hospitality Net*, November 20 Retrieved December 1, 2006, from

19. Kearns, P. 2004. How strategic are you? The six "killer" questions. *Strategic HR Review*, 3(3): 20–23.

20. Kim, B. Y. and Oh, H. 2004. How do hotel firms obtain a competitive advantage?. *International Journal of Contemporary Hospitality Management*, 16(1): 65–71.

21. Mandelbaum, R. 2006. Resort hotels—Wanted: Eaters, golfers, and shoppers. *Hotel Interactive*, June 26 Retrieved June 26, 2006, from http://www.hotelinteractive.com/hi_articles.asp?func-print&article_id=6002

22. Miles, R. E. and Snow, C. C. 1978. *Organizational strategy, structure and process*, New York: McGraw-Hill.

23. Neuendorf, K. A. 2002. *The content analysis guidebook*, Thousand Oaks, CA: Sage.

24. 2005. *Official 2005 North Carolina travel guide*, Raleigh, NC: State of North Carolina Department of Commerce, Division of Tourism, Film, and Sports Development.

25. Olsen, M., West, J. and Tse, E. 2007. *Strategic management in the hospitality industry*, , 3rd Ed., New York: Prentice Hall.

26. O'Neill, J. W. 2006. *Defining segments. Lodging Hospitality* Retrieved September 10, 2006, fromhttp://www.lhonline.com/article/10899

27. Pfeffer, J. 2005. Producing sustainable competitive advantage through the effective management of people. *Academy of Management Executive*, 19(4): 95–106.

28. Perry, C. 1998. Processes of a case study methodology for postgraduate research in marketing. *European Journal of Marketing*, 32(9/10): 785–802.

29. Presbury, R., Fitzgerald, A. and Chapman, R. 2005. Impediments to improvements in service quality in luxury hotels.*Managing Service Quality, 15*, 4: 357–373.

30. Reichel, A. 1983. Strategic management: How to apply it to firms in the hospitality industry. *Service Industries Journal*, 3(3): 329–343.

31. Schiller, P. 2006. *The new luxury* Retrieved January 5, 2005, from http://www.diamonds.net/news/NewsItem.aspx?ArticleID=11089

32. Schuler, R. S. and Jackson, S. E. 2005. A quarter-century review of human resource management in the U.S.: The growth in importance of the international perspective. *Management Review*, 16(1): 11–35.

33. Taylor, M., Finley, D. and Calvert, C. Use of guest workers in seasonal resorts: A pilot study. Proceedings of International Hospitality and Tourism Conference. October. pp.12–15.

34. Taylor, M. and Olsen, M. D. 2006. *Measuring co-alignment in hotels* Unpublished paper

35. Terry, J. W. 2005. Cause for concern: Lodging HR. *American Hotel and Lodging Association*, 6(11)

36. Weinstein, J. 2002. Editor's diary. Defining luxury. *Hotels*, Retrieved September 10, 2006, fromhttp://www.hotelsmag.com/archives/2002/08/editors-luxury-rsuites-baths-restaurants.asp

37. Yin, R. K. 1993. *Applications of case study research*, Newbury Park, CA: Sage.

38. Young, C., McManus, A. and Canale, D. 2005. A value-driven process model of hospitality human capital. *Journal of Human Resources in Hospitality and Tourism*, 4(2): 1–26.

Chapter 3

Historical Evolution of Human Resource Information System (HRIS): An Interface between HR and Computer Technology

Faruk Bhuiyan[1], Mustafa Manir Chowdhury[1], Farzana Ferdous[2]

[1]*Department of Business Administration, International Islamic University Chittagong, Chittagong, Bangladesh*

[2]*Department of Mathematics, University of Chittagong, Chittagong, Bangladesh*

ABSTRACT

Human resources (HR) are the key success factor of any organization operating in the 21st century. From the beginning till today, they have been playing vital role to boost up organizations' competitive capabilities for enjoying competitive advantage and superior profitability over rivals. As today's business environment is more vulnerable due to intense market competition and higher employee empowerment. Organizations have been using information technology for achieving efficiency in human resource management. Thus, it is need for HR professionals to get a clear picture about the historical development of HRM and the increasing role of information technology there in. Consequently, the study tried to analyze critically about evolving human resource information system (HRIS) as a comparatively new technology in HRM. To gratify the stated need, this study analyzed the historical data collected from different published and unpublished sources. The study found that HRIS emerged in the low-cost era (1980-90) and the roles of HR professionals transformed from traditional administrative to more strategic roles during the high-tech era (from 1990).

Keywords: Human Resource Management (HRM), Computer technology, Interface, Historical evolution

1. INTRODUCTION

With the increasing development in technology in the recent years, it is possible to create a real-time information-based, self-service, and interactive work environment. Employee Information Systems have developed from the automated employee record keeping in the 1960s into more complex reporting and decision systems [1]. The last decade observed a remarkable increase in the number of organizations acquiring, storing and analyzing and using human resources data with the help of Human Resource Information Systems (HRIS) [2-5]. Given the authority and relevant accessible information for decision making, human resources (HR) respond more quickly to changes than any time before [6]. The recent emergence of HRIS automates and works out regular administrative and compliance functions that would traditionally perform by corporate HR departments and can facilitate the outsourcing of HR [3]. HR performs traditional HR activities with IT. Since, employees are not just one of the means of production rather a key source of competitive advantage in the 21st century. The authors need to draw a clear picture about theemergence and contributory role of human resources intertwined with developments of Information Technology (IT). Here, a modest attempt made to explore and provide a clear & comprehensive analysis of historical evolution of today's HRIS. The reminding of this paper mentioned here. This paper stated the study method in selection-2 while extensive literature review with authors' own analysis discussed in section-3. Authors evaluated systematic development of HRIS critically in section-4 based on the discussion of section-3. Concluding remarks have drawn in final section.

2. METHOD

Secondary data used that collected from the different journals, research articles, published & unpublished theses, textbooks, company periodicals and websites.

3. LITERATURE REVIEW AND AUTHORS' ANALYSIS

A superficial look into the aims of personnel management showed that this is by far the most important job in industrial management. The materials do not go into the production automatically. Machines are useless without human being to run it. Olivcer Sheldon aptly pointed out that no industry can be efficient so long as the basic fact remains unrecognized that it was the principally human. It is not a mass of machines and technological processes, but a body of man. It is not a complex matter, but a complex of humanity. It fulfills its function not by some impersonal force, but by human energy. Its body is not an intricate maze of mechanical devices but a magnified nervous system. All the activities of an enterprise are initiated and determined by persons who make up that institution. From all the tasks of management, managing human component is the central and most important task because all else depends on how well it works [7]. Prominent Management Gurus concluded that it is not technology, but the excellence of human-and human resource management that forced the continuing challenge for executives in the 21st century [8]. Similarly, it assumes that potential economic and strategic advantage will remain with the organizations those may most effectively attract, develop, and retain a various group of the best and brightest human resources in the marketplace. Historical analysis will demonstrate the growing importance of employees from being just one of the means of production in the 20th-century industrial economy to being a key source of sustainable competitive advantage in the 21st-century knowledge economy. In this study, the historical evolution of HRM had traced out five broad phases of the historical development of industry in the United States [9].

3.1. Stage-I: Pre-world war-ii era and Emerging Personnel Management

3.1.1. Emergence of the Term "Personnel"

Personnel Management (the precursor of the term HRM) can trace back to prehistoric times, when tribal members assigned for specific jobs of hunting or gathering. The households of ancient Chinese emperors had employment tests to identify servants with special talents for special jobs. Then there were the apprentice system and artisan guilds, formed to train new workers. Then human resources treated as "Personnel.

3.1.2. Isolated Function

The then, personnel management isolated from core organizational functions, like; operations, marketing, accounting and finance, and research & development.

3.1.3. The Term "Employee Welfare"

In the late 19th century, where organizations such as; Cadburys at its Bournville factory organized the importance of looking after the welfare of the workforce, and their families. Employing women in their factories in UK during the first world-war leads to introduce "Welfare Officers".

3.1.4. Record Keeping

In the early 20th century and before World War II, the personnel function (the precursor of the term human resource management) primarily involved in record keeping of employee information, like; name, address,, phone, employment history etc. In other words, it fulfilled a "caretaker" function. There was simply no computer technology to automate the records now in history of the course paper records would keep and we can still see paper record Human Resource Systems in many smaller firms today.

3.1.5. Few Govt. Influences

At this point in history, there were few government influences in employment relations, and thus, employment terms, practices, and conditions left to the owners of the firm. As a result, employee abuses such as child labor and unsafe working conditions were common.

Table 1. Summary of pre-world war-ii era

Period	Issues
Pre-world war-ii era	• HR treated as personnel • Personnel management as separate and independent function of organization • Organizations focused on employee welfare programs • Focused on more record keeping about HR • Less emphasized on employee relations

3.2. Stage-II: Post world-war -ii era (1945-60)

The period of 1945-1960 is characterized by the importance of employee morale while personnel are part of operating costs but not yet in mainstream of operations. Research and development in employee selection, payroll automation, applications of mainframe computers for personnel use in defense industry [9] also practiced during this period.

3.2.1. Employee Morale

Labor utilization and mobilization during the war-ii had a great impact on developing the personnel function. Managers realized that employee productivity and motivation had a significant impact on the profitability of the firm. The human relations movement after the war emphasized as employees was motivated not just by money but also by social and psychological factors, such as recognition of work achievements and work norms.

3.2.2. Formal Selection and Development

During the middle of the last century, larger corporations, typically those in the United States that emerge after the second world-war. They recruited personnel from US military and could apply new selection, training, leadership and management development techniques, originally developed by the Armed services, working with, for example, university based occupational psychologists. Similarly, some leading European multinationals, such as; Shell and Phillips developed new approaches to personnel development and drew on similar approaches already used in Civil Service training. Gradually, this spread more sophisticated policies and processes that required more central management via personnel department composed of specialists and generalist team.

3.2.3. Concept of Job Description

Due to the need for classification of large numbers of individuals in military service during the war, systematic efforts began to classify workers around occupational categories to improve recruitment and selection procedures. The central aspect of these classification systems was the job description, which listed the tasks, duties, and responsibilities of any individual who held the job in question. These job description classification systems could also use to design appropriate compensation programs, evaluate individual employee performance, and provide a basis for termination.

3.2.4. Extensive Reporting to Govt. Agencies

Because of the abusive worker practices prior to the War, employees started forming trade unions, which played an important role in bargaining for better employment terms and conditions. There were a significant number of employment laws enacted that allowed the establishment of labor unions and defined their scope in relationship with management. Thus, personnel departments had to assume more record keeping and reporting to governmental agencies.

3.2.5. Usc of Computer Technology

Because of these trends, the personnel department had to establish specialist divisions, such as recruitment, labor relations, training and benefits, and government relations. With its changing and expanding role, personnel departments started keeping increasing numbers and types of employee records, and computer technology emerged for facilitating in maintaining employee information.

Table 2. Summary of post-world war-ii era

Period	Issues
Post World War-II era (1945-60)	• More emphasized on employee relations • Systematic selection • Emerged the term job description for effective selection • Focused on extensive record keeping and reporting • Emerged computer technology as facilitator in managing employee information.

3.3. Stage-III: Legislative era (1960-1980) and Emerging Human Resource Management (HRM)

During the period of 1960-1980, social issues change HR; the increased paperwork and reporting requirements; protector of employees; advent of MIS in computer world; introduction of IBM/360; HR now more in mainstream of operations [9].

3.3.1. Emergence of HR

It was about this time that personnel departments were beginning to be called Human Resources Departments and the field of human resource management was born.

3.3.2. HR Became Key in Organization

The increasing need to be in compliance with numerous employee protection legislations or suffer significant monetary penalties made senior managers aware of the importance of the HRM function. In other words, effective HRM practices were starting to affect the "bottom line" of the firms. So, there was a significant growth of HR departments, and computer technology had advanced to the point where it was beginning to use.

3.3.3. Govt. and Regulatory Agencies Increased Reporting Requirement

This period witnessed an unprecedented increase in labor legislation that governed various parts of the employment relationship, such as prohibition of discriminatory practices (EEO), occupational health and safety (OSHA), retirement benefits, and tax regulation. As a result, the HR department burdened with the additional responsibility of legislative compliance that required collection, analysis, and reporting of voluminous data to statutory authorities.

3.3.4. Development of MIS for HRM

There was an increasing demand for HR departments to adopt computer technology to process employee information more effectively and efficiently. This trend resulted in an explosion in the number of vendors who could assist HR departments in automating their programs in terms of both hardware and software. Simultaneously, computer technology was evolving, and delivering better productivity at lower costs. These technology developments and increased vendor activity led to the development of a comprehensive management information system (MIS) for HRM.

Table 3. Summary of legislative era

Period	Issues
Legislative Era (1960-1980)	• Personnel became HR • HR treated as key for organization success • Increased govt. pressure on proper report keeping • Emergence of MIS for HRM • HRIS used mostly to keep administrative records.

3.3.5. Limited use of HRIS

The decreasing costs of computer technology versus the increasing costs of employee compensation and benefits acquire of computer-based HR systems (HRIS) a necessary business decision. However, the personnel departments were still slow in adopting computer technology, even though it was inexpensive relative to the power it could deliver for the storage and retrieval of employee information in MIS reports. Still, HRIS would use mostly for keeping administrative records.

3.4. Stage-IV: Low Cost Era and Emerging HRIS (1980-1990)

With increasing competition from emerging European and Asian economies, U.S. and other multinational firms increased their focus on cost reduction through automation and other productivity improvement measures. The increased administrative burden intensified the need to fulfill a growing number of legislative requirements, while the overall functional focus shifted from employee administration to employee development and involvement. To improve effectiveness and efficiency in-service delivery, through cost reduction and value-added services, the HR departments came under pressure to harness technology that was becoming cheaper and more powerful and desired cost-effective HR software like HRIS. The history of HRIS began from payroll systems in the late 1950s and continued into the 1960s when the first automated employee data used [9].

3.4.1. Affordability of HRIS Regardless of the Size of the Organization

In the 1960's and 1970's, large companies felt a need to centralize their personnel data in large part to facilitate record keeping and meet regulatory needs. Programs would write on large mainframe computers that acted as a central data repository with little transactional processing, usually only for payroll. The Human Resource Information System (HRIS), also known as a Human Resource Management System (HRMS), became prevalent in the 1980's with the popularity of Enterprise Resource Management (ERP) applications and the move from mainframe systems to client server technology.

3.4.2. HRIS as Simple Record Keeping to Complex Analytical Tools

HRIS has evolved from simple record keeping to complex analytical tools to assist management decision making. The 1980's saw a shortage in skilled workers, especially in the technology sector. Human Resource Management had long evolved from the basis of a skills management discipline to more of an employee satisfaction and productivity tool. However, by the 1980's, HRIS systems now included a host of feature sets and functional capabilities aimed at attracting, retaining and properly compensating the workforce. By 2000, the human resource software industry saw HRIS grow to include recruitment, benefits management, time management, payroll, compensation management, learning management, expense reporting and reimbursements, and performance management. Self-service applications built on top of the underlining data empowered employees to manage their own data and make timely changes. Online employee portals further consolidated disparate systems, documents and information into one place.

Table 4. Summary of low cost era and emergence of HRIS

Period	Issues
Low Cost Era and Emergence of HRIS (1980-1990)	• HRIS became affordable to organizations • HRIS became complex analytical tool • HRIS for all

Table 5. Differentiating features of both legislative and low cost era

Basis	Legislative Era	Low Cost Era
Time	1963-1980	1980-1990
Focus	HR assumed key	HRD concentrates on HRIS
Regulators	increased reporting requirements by Govt. and Regulatory Agencies	Increased demand for ensuring a technology-enabled HRM system by giant competitors.
Scope	HRIS used mostly to keep administrative records.	HRIS as simple record keeping to complex analytical tools.

Table 6. Historical Highlights of Humanic offering

Year	Name of HRIS service providers	Solutions Offered
1982	Humanic Design	Empire HRMS (was DOS based)
1987	Humanic Design	Rational Database Product
1991	Fiserv	Fiserv HRIS
1994	Humanic Design	HRMS solutions in the U.S. (Windows based)
1999	Fiserv & Humanic Design	Humanic Solutions
2000	Humanic	Integrated HRMS/Payroll/self-Service Offering with Drag and Drop reporting
2000	Humanic	Application Service Provider(ASP) Solution

3.4.3. HRIS for Large and Small Businesses

As computer hardware prices fell and computing power simultaneously grew, more and more companies could afford enterprise software systems and vendors saw a market for standalone HRIS software. Data connectors and application programming interfaces empowered customers with HR systems that need not delivered with their financial accounting software. They now have an a la carte option and can leverage a higher fit system that better fulfills their HRIS needs and can integrate information such as payroll and headcount to their financial system.

Humanic (formed as result of merger between Fiserv HRIS, founded in 1984 and Humanic Design Corporation, founded in 1982) solved many HR, Payroll, benefits and challenges. The following table shows the historical highlights of Humanic greatest strength was in its in-depth knowledge and experience of its professional staff in all facets of the HRMS/Payroll industry.

3.5. Stage-V: Tech Era and Emergence of Strategic HRM (1990 to the present)

The increased use of technology and changed focus of the HRM function as adding value to the organization's product or service led to emerge the HR department as a strategic partner. With the growing importance and recognition of the people and people management in contemporary organizations, SHRM has become critical in management thinking and practice. SHRM derives it theoretical significance from resourced based view of the firm that treats human capital as a strategic asset and a competitive advantage in improving organizational performance [10]. SHRM designed to diagnose firm strategic needs and planned talent development, which requires implementing a competitive strategy and achieving operational goals.

Strategic Human resource Management (SHRM) evolved from Personnel Management in two-phased transformation: from personnel management to traditional human resource management (THRM) and then from THRM to SHRM [11]. Firms today realize that innovative and creative employees who hold the key to organizational knowledge provide a sustainable competitive advantage because unlike other resources, intellectual capital is difficult to imitate by competitors.

4. CRITICAL EVALUATION

Many researchers have tried to explore the true picture of emerging the computer aided HRM (e-HRM) from longtime before; still today, it is a continuous attempt to many researchers. Findings of the earlier researchers had published in different journals. Based on the literature survey, the study found that few articles showed the real, more comprehensive and easily understandable outcomes on the chronological emergence of today's HRIS. Consequently, here it has tried to explore the clear, almost complete and coherent summary on the chronological development of HRIS. This has done based on the five development stages of U.S. industry.

During the pre-world war-ii, the term personnel management mostly used and labor would treat as personnel. Personnel management evolved as an isolated function as other core functions of organization and the term employee welfare would use. On the other hand, government had few influences on HR policies, rules, regulations regarding personnel. Since, owners were free from government interference, child labor and unsafe working conditions were common. The personnel function primarily involved in record keeping of employee information as course paper because there was simply no computer technology to automate the records now.

After the war, it observed that employees motivated not just by money but also by social and psychological factors, such as recognition of work achievements and work norms. There were new selection, training, leadership and management development techniques, originally developed by the Armed Services working with university based occupational psychologists. The job description classification systems could also use to design appropriate compensation programs, evaluate individual employee performance, and provide a basis for termination. Since personnel would deprive from their rights and privileges, they formed trade union and organizations were under more pressure than pre-world war-ii to keep more record and report to the government agencies. As personnel department characterized by more and wider functions, computer technology began to emerge as a possible way to store and retrieve employee information.

During the period of legislative era (960-1980), the term personnel and personnel management replaced by the term human resource and human resource management. On the other hand, the HR department burdened with the additional responsibility of legislative compliance that required collection, analysis, and reporting of voluminous data to statutory authorities. There was an increasing demand for HR departments to adopt computer technology to process employee information more effectively and efficiently. These technological developments and increased vendor activity led to develop a comprehensive management information system (MIS) for HRM. Though the decreasing costs of computer technology versus the increasing costs of employee compensation and benefits motivate acquiring of computer-based HR systems (HRIS) a necessary business decision yet HRIS used mostly for keeping administrative records.

Since, HR departments approached under pressure to link with technology that was becoming cheaper and more powerful and desired cost-effective HR software during low cost era (1980-1990). HRIS became prevalent in the 1980's to meet the then HR demand with the popularity of Enterprise Resource Management (ERP) applications and the move from mainframe systems to client server technology. In addition, HR department with the upgrading information technology evolved from simple record keeping to complex analytical tools to assist management decision making. It not only evolved for larger organizations with huge capital but also for small enterprises with low capital. In fact, HRIS applications in true sense observed during this period (1980-1990).

The drastic change in the information technology altered the role of HR professionals from traditional to strategic at the last decade. HR professionals with computer aided HRM are now in the position to be more strategic to contribute in gaining competitive advantage, improving organizational overall performance and better knowledge management. Hence, HRIS is a predictor of strategic human resource management (SHRM).

In fine, the study found the use of personnel management before world war-ii, traditional HRM during the legislative era (1963-1980), computer aided HRM (e-HRM) during the period of low cost era (1980-1990). Further, it observed the roles of HR professionals have transformed from traditional HRM to Strategic HRM (SHRM) during high-tech era (from 1990). In this junction, HRIS plays vital role to the changing roles (strategic) of HR professionals so they can be more responsive to the vulnerable and intense competitive market. HR professionals have now specific HR agenda about how they to be more strategic with the efficient application of HRIS. The following figure shows the sequential evolution of HRIS and SHRM.

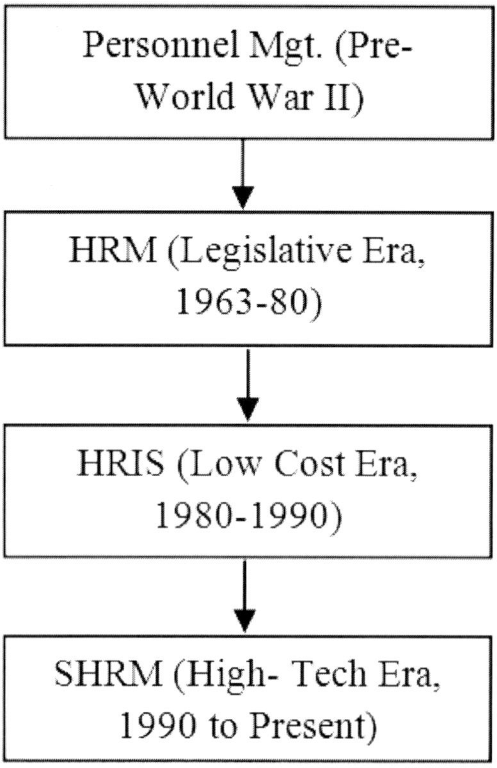

Figure 1. A Sequential Evolution of HRIS and SHRM

5. CONCLUDING REMARKS

Companies in the 21st century are under pressure to reduce the cost of the production and to be more responsive to the customers. Moreover, the success of today's companies mostly depends on the effective and efficient utilization of their human resources. Since, HR professionals not only play administrative but also strategic roles with the advent of IT-enabled HRM, HR managers should have clear idea about the background of HRIS. This study showed the true picture of HRIS evolution based on theoretical analysis. Hence, Personnel Mgt. HRM, HRIS, and SHRM emerged during the pre-world war, legislative era (1963-80), low cost era (1980), and high-tech era (1990 to Present) respectively. The outcomes of the study will assist HR professionals, HR managers, HR executives, employees, owners, government, and regulatory bodies to formulate HR related policies for ensuring the proper application of HRIS. This study will also support new researchers to study advance regarding this matter. Finally, this study suggests researchers for further study, to gather more realistic knowledge about today's HRIS, by collecting information directly from different HR professionals, management gurus and successful entrepreneurs with the published sources of information.

REFERENCES

1. Gerardine DeSanctis, *Human Resource Information Systems- A Current Assessment.* MIS Quarterly, 1986. 10(1): p. 15-27.
2. Ball, K.S., *The use of human resource information systems: a survey.* Personnel Review, 2001. 30(5/6): p. 677-93.
3. Barron M., et al., *Exclusive Panel Discussion: Tips and Trends in HRIS.* HR Focus, 2004. 81(5): p. 6-7.
4. Hussain Z., Wallace J., and Cornelius N. E., *The use and impact of human resource information systems on human resource management professionals.* Information & Management, 2007. 44(1): p. 74-89.
5. Ngai, E.W.T. and F.K.T. Wat, *Human resource information systems: a review and empirical analysis.* Personnel Review, 2006. 35(3): p. 297-314.
6. Lengnick H. M. and Lengnick H. C., *New Challenge; New Roles; New Capabilities* in *Human Resource Management in the Knowledge Economy.* 2002, Berrett-Koehler: San Francisco.

7. G. Biswanath, *Personnel Management, Industrial Relations and Labor Welfare*. The Millennium Edition ed. 2000, Calcutta: World Press Private Limited

8. Drucker P. F., et al., *Looking ahead: Implications of the present.* Harvard Business Review, 1997. 75(5): p. 18-24.

9. Kavanagh M. J., Gueutal, and Tannenbaum S. I., *Human Resource Information Systems*. 1990: Boston: PWS-Kent.

10. Becker, B.E. and M.A. Huselid, *Strategic human resource management: Where do we go from here?* Journal of Management, 2006. 32(6): p. 898-925.

11. Schuler, R.S., S. Dolan, and S.E. Jackson, *Introduction to strategic human resource management.* International journal of manpower, 2001. 22: p. 1995-7.

Chapter 4

Why Should Gender Differences In Hospitality Really Matter? A Study of Personnel's Service Orientation and Job Satisfaction in Hotels

Marko D. Petrović[a]*, Tamara Jovanović[a], Jelica J. Marković[a], Tanja Armenski[a] & Vladimir Marković[a]

Department for Geography, Tourism and Hotel Management, Faculty of Sciences, University of Novi Sad, Novi Sad, Serbia

ABSTRACT

In this article we examine the gender differences among hotel employees in service orientation and job satisfaction. To determine these differences, we used a service orientation scale (SOS), developed by Dienhart, Gregoire, Downey and Knight and a job satisfaction scale developed by Lytle. Our assumptions were that there would be significant gender differences in service orientation and job satisfaction. Our research proved factor structures of the two scales we used. Specifically, we showed gender differences in two of three factors, concerning service orientation (organisational support and customer focus), and no gender differences in job satisfaction. Our study revealed that organisational support is more important to men because they value more structured labour and service procedures than women ($t = 2.21$, df = 110, $p = 0.03$). In contrast, customer focus is more relevant to a women because it is more crucial to see satisfied guests, as well as having a good relationship with them ($t = 2.07$, df = 110, $p = 0.04$). It is of great importance for hotel management to have in mind that male and female employees respond and behave differently depending on the stimuli in the work environment and their satisfaction varies because of it.

Keywords: gender differences, service orientation, job satisfaction, hotel employees, Novi Sad, Serbia, C10, L83, O52

1. INTRODUCTION

Hotel enterprise, as an important part of hospitality, represents combination of spatial and functional service of accommodation, including food and beverages services as well as all the other supporting services. In recent years, human resources (HR) management in hospitality and tourism, together with proper employee selection, is becoming increasingly important for the success of hotel enterprises (Bolton & Boyd, 2003; Cheng & Krumwiede, 2012; Chiang & Birtch, 2011; Hoque, 2000; Kim, Leonga, & Lee, 2005). Therefore, employees are becoming the centre of interest of tourism academics and practitioners. Considering this, together with the fact that hotel industry implies working with people, studying hospitality concepts, such as *service orientation* and *job satisfaction* of the employees, becomes essential.

Having gone through many phases of transformation of organisational components, today's hotel industry has significantly changed in the gender structure of employees. The number of women employed in this type of business rises (Purcell, 1996; Iverson, 2000; Pinar, McCuddy, Birkan, & Kozak, 2011). Therefore, women are employed at all levels of the organisational structure, including management, and this brings changes in the general conception of the hotel business (Iverson, 2000; Aykac, 2006; Cave & Kilic, 2010; Pinar et al., 2011; Campos-Soria, Marchante-Mera, & Ropero-García, 2011; García-Pozo, Campos-Soria, Sánchez-Ollero, & Marchante-Lara, 2012).

A similar situation occurs in Serbia. Hotel companies in Serbia are going through transitional changes that affect the business in general as well as the relationship of the company towards employees. These changes cause already detectable differences in Serbian hospitality that have not yet been thoroughly examined by the scientific community (Nedeljković, 2008). Therefore, we opted to investigate gender-conditioned differences in hotels in Novi Sad, with the focus on two hospitality variables: service orientation and job satisfaction.

2. SERVICE ORIENTATION OF EMPLOYEES

Service orientation of employees could be defined as the ability to be useful, thoughtful, considerate and cooperative in providing services at the individual level (Dienhart, Gregoire, Downey, & Knight, 1992). On the other hand, authors such as Lytle, Hom, and Mokwa (1998, p. 459) defined service orientation as 'an organisation-wide embracement of a basic set of relatively enduring organisational policies, practices and procedures intended to support and reward service-giving behaviours that create and deliver service excellence'.

Service orientation of employees is determined by many factors – from the work environment, job demands, and interpersonal relationships within the organisation to financial and personal satisfaction. Much research has showed that during the use of certain hotel services, employee treatment of customers is the most important issue (Galičić & Ivanović, 2008; Hartline & Ferrell, 1996; Nikolić, 2006; Petrović & Marković, 2011). The consequences of service orientation are expressed in terms of job satisfaction of the employee (Jaworski & Kohli, 1993; Lee, Park, & Yoo, 1999), and the turnover that employees make (Lahey, 1984).

Moreover, some research shows the role of service orientation in strengthening competitive advantages (Dienhart, Gregoire, Downey, & Knight, 1992; Schneider & Bowen, 1985; Kim, Leonga, & Lee, 2005; Kim, 2008). The authors considered that employees should have the competence to be cooperative and thoughtful in providing hotel services at a personal level, and most importantly, to have individual access to each guest in the hotel, so they can provide quality service. Employees in the hotel industry should have adequate knowledge and skills in order to provide not only the expected quality, but also additional quality service to the guests, as they are definitely the most important issue when evaluating hotel service quality. Numerous studies have shown that kindness and accessibility of staff are more important to guests than competence and technical organisation (Čerović, 2002; Galičić & Ivanović, 2008). The attitudes and behaviour of employees results in the perception of service quality by customers (Hartline & Ferrell, 1996). Establishing relationships with service users is crucial to building loyalty and improving the service delivery process (Galičić & Ivanović, 2008; Nikolić, 2006). Certain psychological characteristics (sociability, openness, agreeableness, etc.) enable employees to meet with the increasingly complex demands of the travel industry.

In the field of tourism and hospitality, results of Dienhart, Gregoire, Downey, and Knight (1992) suggest that service orientation is comprised of three components: *organisational support, customer focus,* and *service under pressure*, but also suggest that increasing employees' job satisfaction (with job involvement and job security) is involved in the improving of their overall service orientation. They conclude that there are positive relationships between job satisfaction and customer focus, employees' favourable perceptions of job involvement and job security.

Thirteen years later, Kim and colleagues (Kim, Leonga, & Lee, 2005) used Dienhart et al.'s research (1992) to examine the relationship between hospitality employees' service orientation and their job satisfaction, organisational commitment and intentions to leave. They found that only two elements of service orientation (organisational support and customer focus) have a significant positive correlation with job satisfaction. The third element, service under pressure, has no correlation with the observed variable.

The role of emotions in the service orientation is crucial. Hochschild (1983) argued that service agents (e.g. hotel employees) are expected to experience and express certain feelings during service interactions. The concept of service orientation implies that workers should be able to display all the emotions that are desired by their supervisors and consumers (O'Connor, Trinh, & Shewchuk, 2000). Emotional labour is defined as a display of expected emotions by service agents during service encounters, and since service context involves face-to-face interactions it is important to explore the emotions and emotional labour in a service orientation context (Ashforth & Humphrey, 1993). Hochschild (1983) claimed that service providers comply with expression norms through surface acting, self-expression, and expression of spontaneous and genuine emotion. However, the presence of a service agent, such as a hotel employee, may prime expectations of good service that cannot be met, and emotional labour may trigger emotive dissonance and impair one's sense of authentic self.

Few studies have dealt with gender differences in service orientation and those that have, showed inconclusive findings. Study of Kim, McCahon, and Miller (2003) show the effect of employee characteristics (gender, age, marriage, education, etc.) on service orientation in restaurants in Seoul, South Korea. They found that employees with a longer length of service and those in supervisory positions display a higher degree of service orientation, whereas other demographic data such as gender do not have a significant impact on employee service orientation. Finding of Kim, McCahon, and Miller (2003) contradicts the study of Groves (1992) who found gender differences in service orientation. According to his research, females had more of a*personal focus* on service, and they perceived more *support* from the organisation, and viewed service as less stressful *under pressure* than male subjects.

Gender could affect service orientation in terms of individual psychological characteristics and issues of emotion management. Many researches have explored gender differences in emotional labour and emotional control in service procedures (O'Connor, Trinh, & Shewchuk, 2000). Rafaeli (1989) claims that women display more positive emotions to customers than men, and she denotes three reasons why: (1) higher ability to empathise and show their emotions (LaFrance & Banaji, 1992); (2) women are brought up to behave themselves in a pleasant and agreeable fashion (Deutsch, 1990; James,1989); and (3) women need extra social approval (Hoffman, 1972).

Studies in Western societies indicate that there are many gender stereotypes, such as the one that women are free to express emotions and men are taught to control these emotions. Research by Leidner (1993) and Carmeli (2003) have shown that women are more likely to express empathy, pleasantness and helpfulness to a greater extent than their male colleagues. Caruso, Mayer, and Salovey (2002) points out that women have a great ability to understand other people's emotions and are closer in communication with others. There is research that show that women are better at controlling emotions of stress at the workplace (Kruml & Geddes, 2000). They have a higher frequency of expression of desired emotions, even when they are employed on the same job as their male colleagues. However, research shows that women do not show greater sensitivity to stress as a consequence of emotional labour (Bandura, 1982; Kruml & Geddes, 2000; Morris & Feldman, 1997; Wharton, 1999; Wharton & Erickson, 1993), and although it is assumed that women are frequently employed on the jobs that require emotional work, both genders express the same degree of negative consequences.

3. JOB SATISFACTION OF EMPLOYEES

The second important issue of our research is personnel's job satisfaction in the observed hotels. Odom, Boxx, and Dunn (1990, p. 159), defined job satisfaction as 'the extent to which a worker feels positively or negatively about his or her job' and Lytle's (1994) research indicates that job satisfaction is a unidimensional variable. Job satisfaction is an important predictor of quality of hotel services. There are many determinants of job satisfaction and certain research even suggests that job satisfaction may be genetically determined and remains constant over a five years period though many segments of business change (Arvey, Bouchard, Segal, & Abraham, 1989; George, 2000). However, we do not believe that job satisfaction is a purely genetically-determined construct but is influenced by many other factors: motivation, professional environment, spouse, family, friends, etc.

In a service business, job satisfaction ensures that employees treat service users, i.e. hotel guests/tourists with maximum respect (Arnett, Laverie, & McLane, 2002). Because of the importance of contact with service users due to the development of relations with them, the employee's job satisfaction is a major concern for service companies, i.e. hotels that want to keep a large number of loyal customers. Other authors, such as Jerome and Kleiner (1995), agree that enterprises which are working to improve employee's job satisfaction, motivation and job dedication, can expect long-term benefits of corporate success, loyalty and productivity of employees. Some authors claim that job satisfaction also has a positive correlation with a commitment to organisations (Blešić, Popov-Raljić, & Romelić, 2007; Blešić, Ivkov-Džigurski, Dragin, Ivanović, & Pantelić, 2011; Humphreys & Williams, 1996) and performance (Birnbaum & Somers, 1993; Babin & Boles, 1996). Shaw (1999) showed that there is a strong negative relationship between the level of turnover and job dissatisfaction. Others believe that if employees are dissatisfied with their job, it is more likely that they will make no expectations and eventually leave the organisation (Mobley, 1977).

There are many models (Hunt & Chonko, 1984; Lawler & Suttle, 1973; Lucas, Parasuraman, Davis, & Enis, 1987; Teas, 1983; Sujan, 1986) which create satisfied (hotel) employees and improved quality of operations. Increase in the responsibility of the employee creates a greater commitment and sense of usefulness and importance of the employee and the organisation. More power leads to a timely resolution of unforeseen problems in working with guests, but is also creates satisfaction. Employees should be aware of changes and innovations in business enterprises such as hotels.

The relationship between job satisfaction and gender differences has been examined many times, with contradictory results. Some studies have shown women to be more satisfied than men (Clark, 1996, 1997; Ishitani, 2010; Sloane & Williams, 1996), whereas others have shown men to be more satisfied than women (Aydin, Uysal, & Sarier, 2012; Kim, Murrmann, & Lee, 2009; Mora & Ferrer-i-Carbonell, 2009). However, a number of the researchers reported no significant differences between genders in relation to job satisfaction (De Vaus & McAllister, 1991; Smith, Smits, & Hoy, 1998; Sousa-Poza & Sousa-Poza, 2007; Ugorji,1997).

The research of De Vaus and McAllister (1991) shows gender differences in regard to what men and women look for in a job. According to Malka and Chatman (2003) job orientation represents work-related preferences to value specific types of rewards inherent in the work environment. Some other studies, such as Loscocco (1989), show that women value extrinsic job characteristics more than men do. However, there are other researches that show the opposite (Neil & Snizek, 1987). Finally, certain studies have found no differences between men and women in terms of intrinsic and extrinsic job orientation (Brief, Rose, & Aldag, 1977). Some authors examine gender differences in job satisfaction and service orientation, using closely comparable data collected in nine Western European countries (De Vaus & McAllister, 1991). Their results show that men place greater value than women on both extrinsic and intrinsic work values and are somewhat more satisfied than women with their jobs.

4. THE PRESENT STUDY

Most of the research on gender differences of service orientation and job satisfaction was conducted in the western societies, and the study results are far from conclusive. Thus, gender distinctions in hospitality have not yet been resolved. To the best of our knowledge, the influence of gender differences on service orientation and job satisfaction has not been studied in previous research in our region (south-eastern Europe). The same goes for Service Orientation Scale (SOS) developed by Dienhart et al. (1992), which will be explained in the Method section.

The purpose of our study was to investigate gender distinctions in service orientation and job satisfaction in order to identify some differences and similarities that are conditioned by the behaviour of men and women. This research will provide insights into gender differences of two important variables in providing quality service in the hospitality industry in Novi Sad, Serbia.

Given such research is scarce in the Balkans in general, this research might serve to stimulate similar surveys in the future. Also, the present research is important because it is one of the first empirical examinations of the gender influence on job satisfaction and service orientation in hotel industry in our region. Moreover, Serbian society is currently undergoing a period of adaptation to new labour market changes and women are occupying an increasingly important role in the contemporary work processes in hotel business. According to the Statistical Office of the Republic of Serbia in 1992, the gender structure within hotel employees shows were only 28% of hospitality employees, and in 2012, women dominate in this kind of business, being 52% of all hospitality employees (http://webrzs.stat.gov.rs/WebSite/). Therefore, there is a significant change in the gender structure in hotel industry, but also in perception of the job and relationship with it.

Considering the topic of this study, we expect significant gender differences in service orientation and job satisfaction, because men and women show quite dissimilar behaviour in these two variables (Groves, 1992; Kim, McCahon, & Miller,2003). According to these findings, we made a few assumptions:

- women in the observed hotels show more customer focus than men, because women are more motivated to provide more quality and faster service, in order to prove themselves in this business.
- it is more important to a woman to see satisfied clients, i.e. guests than a support and encouragement of the managers, because we believe the internal motivation is more crucial to a female than a male employee.
- women will perceive more work situations under pressure.
- men would value and perceive more organisational support and precise work procedure, than women.
- we assume that women will show higher level of job satisfaction than men, like in the studies mentioned earlier (Clark,1996; 1997; Hodson, 1989; Murray & Atkinson, 1981; Sloane & Williams, 1996).

5. RESEARCH METHODOLOGY

5.1. Research sample

The population for this study consists of 112 employees (sample size) of 16 (city type) hotels in Novi Sad ('Aurora', 'Park', 'Leopold I', 'Zenit', 'Mediteraneo', 'Gimnas', 'Centar', 'Duga', 'Best Western Prezident', 'Elite', 'Vojvodina', 'Putnik', 'Novi Sad', 'ARTA boutique', 'Vigour' and 'Panorama'). Sampling was convenient since subjects were employees willing to participate in the study. Employees were selected from eight segments of the hotels' organisational structure: Front Office/Reception, HR Management, Restaurant, Bars, Administration and Accounts, Housekeeping, Executive Management and Kitchen. Each work position was approximately represented, with almost equal number of respondents for each category. While sampling, we covered all categorisations levels of hotels (from one star to five star hotels). Within our study we investigated the information-seeking of male and female employees and employers in observed hotel enterprises.

Table 1 presents the distribution of respondents' sex, work position, age, educational background and residence. On the basis of survey research it can be concluded that among the employees a significantly higher number were female (nearly 59%), 41% of respondents were men. The ratio is almost the same as for the entire population of travel industry in Serbia of whom 56% are women and 44% are men (Čačić, 1998; Čerović, 2002).

Table 1. Sample characteristics (%).

	Men	Women
Total:	**41.1**	**58.9**
Work Position:		
Employer	22.5	14.7
Staff	77.5	85.3
Age:		
less than 20 years	11.3	14.8
21–30 years	30.2	32.3
31–40 years	21.1	20.8
41–50 years	13.2	24.1
51–60 years	18.2	5.7
61+ years	6.0	2.3
Education:		
Elementary School	10.3	6.1
High School	45.7	41.6
BSc Degree	36.2	47.0
MSc/PhD Degree	7.8	5.3
Residence:		
City of Novi Sad	75.1	82.8
Surrounding area	22.4	16.8
Other settlements	2.5	0.4

Source: Data gained in the research.

Among the respondents, nearly 80% reside in the territory of Novi Sad (men 75%, women 83%) and the remainder lives in surrounding smaller towns (men 22%, women 17%). Only a few respondents are from the other regions in the country (men 2.5%, women 0.4%). Most (30.2%) of the male respondents were in their 20s; the next largest group (21.1%) of men were in their 30s. Most (43.8%) of the women were also in their 20s and 24.1% were in their 40s. When analysing the structure of respondents by working position, results clearly show higher proportion of employees in the executive affairs (men 77.5%, women 85.3%) than employees in managerial positions (men 22.5%, women 14.7%). Further results present the educational structure and clearly show the highest percentage of respondents with BSc diploma, and slightly less with a secondary school diploma. Most male respondents (56%) have elementary and high school diplomas; 44% had a BSc and MSc/PhD diploma. Most female respondents (52.3%) have elementary and high school diplomas; 47.7% of the respondents had a BSc and MSc/PhD diploma.

5.2. Research instruments

The questionnaire consists of three parts. The first part involves items which concerned socio-demographic characteristics of respondents (age, gender, work position, level of education, etc.) (Table 1). Second part consists of the list of nine items about the *service orientation* of respondents (Table 2). This questionnaire is a translation of the SOS developed by Dienhart et al. (1992). This scale measures three dimensions of service orientation: customer focus, organisational support and service under pressure (Kim, Leonga & Lee, 2005). Items 1, 2, and 6 of the SOS, should represent *customer focus. Organisational support* is described with items 3, 4, 5 and 8. Finally, items 7 and 9 should depict *service under pressure*. The third part shows the list of five items about the *job satisfaction* of the employees (Table 3). The job satisfaction scale was adapted from Lytle's research (1994), with only five items (Kim, Leonga & Lee, 2005). We decided to make our study more functional, so we used an abbreviated version of the questionnaire, where we included five items which fit into planned research. In both scales, the respondents express their opinion about items on the 5-point Likert scale (from 1 – completely agree to 5 – completely disagree). Both scales were back translated to ensure proper measurement.

5.3. Research hypotheses

We examined relationships between service orientation (customer focus, organisational support, and customer service under pressure), job satisfaction and gender. Based on the literature review, several hypotheses were formulated:

Table 2. Service orientation (SO) of hotel employees.

SO1	Interacting with customers is enjoyable
SO2	It is important to me that the customer is satisfied
SO3	The employees at my hotel provide excellent service
SO4	My manager encourages me to provide better service
SO5	The training which I received prepared me to provide excellent service
SO6	Customers treat me with respect
SO7	At peak hours, we are so busy that we cannot provide excellent service
SO8	Our service procedures make it easy for me to give excellent service
SO9	My manager expects us to always follow procedures, even if it means giving less than excellent service

Source: Dienhart, Gregoire, Downey, and Knight (1992), in Kim, Leonga, and Lee (2005, p. 178).

Table 3. Job satisfaction (JS) of hotel Employees.

JS1	I consider my job pleasant
JS2	I feel fairly-well satisfied with my present job
JS3	I definitely like my work
JS4	My job is pretty interesting
JS5	I find real enjoyment in my work

Source: Lytle (1994), in Kim, Leonga and Lee (2005, p. 179).

Hypothesis 1: *Customer focus is positively correlated with job satisfaction.*

Hypothesis 2: *Organisational support is positively correlated with job satisfaction.*

Hypothesis 3: *Customer service under pressure is negatively correlated with job satisfaction* (Kim, Leong, & Lee, 2005).

Hypothesis 4: *Women show more customer focus than men.*

Hypothesis 5: *Men would perceive more organisational support than women.*

Hypothesis 6: *Women would perceive more service under pressure.*

Hypothesis 7: *Women are more satisfied with their jobs than men.*

Based on the proposed hypothesis, the model of relationship between examined variables is shown in Figure 1. The arrows in the Figure 1 show that examined variable (gender) has impact on service orientation and job satisfaction, as well as those two variables have impact on each other. On the other hand, the lines mean that service orientation consists of three dimensions: customer focus, organisational support and service under pressure. Figure 2 illustrates significant gender differences in organizational support and customer focus. According to the figure, we can notice that organizational support is more important to men, because men value more structured labour and service procedures than women (t = 2.21, df = 110, p = 0.03). On the other hand, customer focus is more relevant to a women, because it is more crucial for them to see satisfied clients that respect them ($t = 2.07$, df = 110, $p = 0.04$).

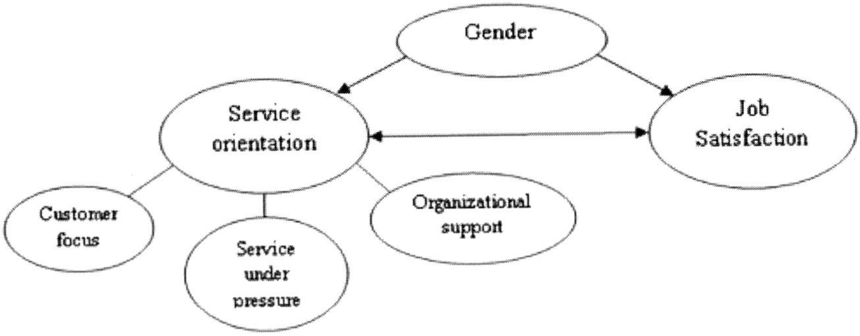

Figure 1. A proposed model of relationships between variables.

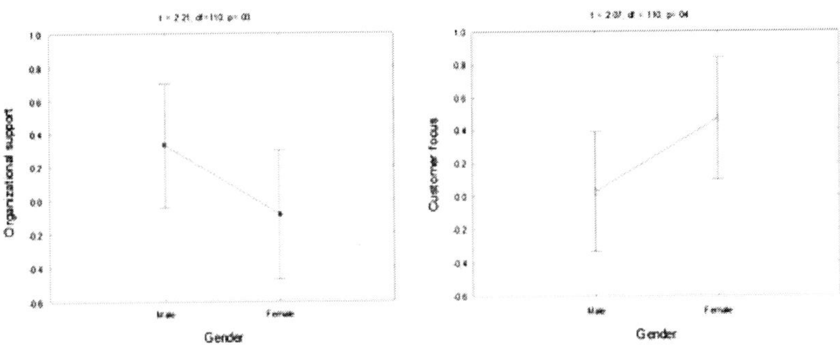

Figure 2. Gender differences in organisational support and customer focus.

5.4. Research procedure

The research was carried out from March until June 2012. We distributed the questionnaire to all employees who expressed an interest in taking part in this research. At the beginning, we informed the respondents about the subject of examination. We distributed the questionnaires in respondent's place of work and then came after a few days to pick them up. In order to ensure respondents' anonymity, we undertook the following: emphasised that they will not write their names on the questionnaires, they could take questionnaire to their home and, after filling them out, they should return questionnaires via another employee (for example, the receptionist), who returned the questionnaires to the researchers.

6. RESEARCH RESULTS

Before we have analysed the data, we have recoded all the answers to the items which were not in the same direction as the rest (SO7 and SO9). Firstly, we wanted to confirm latent structures of our two scales: *job satisfaction* (unidimensional) and*service orientation* (three-dimensional). Principal component factor analysis with Varimax rotation on total sample was used to identify these underlying dimensions.

Using factor analysis, it was found that job satisfaction has unidimensional structure with scale reliability coefficient of $\alpha = 0.89$, so we simply summed the answers in the form of a composite score. On the other hand, we extracted three factors with eigenvalues higher that 1, which describe service orientation (see Table 4):

- Factor 1: *organisational support,*
- Factor 2: *customer focus and*
- Factor 3: *service under pressure.*

Table 4. Service orientation (orthogonal rotation).

	Eigenvalue	% Total Variance
Factor 1	3,032	33,687
Factor 2	1,362	15,132
Factor 3	1,052	11,689

Source: Data gained in the research.

Factor 1 explains about 34% of the total variance, Factor 2 explains additional 15% of the variance, while the Factor 3 covers nearly 12%. The total percentage of explained variance is about 60%.

Kaiser-Meyer-Olkin measure of the sampling adequacy is KMO = 0.71 and Bartlett's test of sphericity is $\chi^2_{(36)} = 229.147$, significant at $p < 0.01$, therefore showing that variables correlate between themselves significantly.

Based on Table 5 we can conclude that the highest number of items represents Factor 1 which we concluded resembles dimension *organisational support* (Dienhart, Gregoire, Downey, & Knight, 1992). Item SO8 has the highest positive correlation with Factor 1 (factor loading = 0.794). This means that the establishment of service procedures that facilitates the provision of excellent hotel services is highly correlated with the organisational support. We expected this finding in our initial hypothesis. Variable SO4 also highly correlated with Factor 1, which clearly shows that encouragement by managers in providing quality services is an important part of organisational support. On the other side,

SO6 loads onto Factor 1 as well, which leads to an unusual conclusion where a large number of the respondents believe that respect from the service consumers, i.e. hotel guests, leads to better organisational support in the hotel company. Therefore, certain contradiction arises in relation to the definition of organisational support, which basically does not include relationship with guests, but only internal relationships in organisation, i.e. 'the degree to which supervisors value the contributions of employees and care about their well-being' (Eisenberger, Huntington, Hutchison, & Sowa, 1986; Kottke & Sharafinski, 1988; Rhoades & Eisenberger, 2002).

Table 5. Factor analysis – service orientation.

		Factor loadings	Reliability coefficient
	Factor 1		0.73
SO8	Our service procedures make it easy for me to give excellent service	0.794	
SO4	My manager encourages me to provide better service	0.773	
SO6	Customers treat me with respect	0.708	
	Factor 2		0.64
SO2	It is important to me that the customer is satisfied	0.868	
SO1	Interacting with customers is enjoyable	0.663	
SO3	The employees at my hotel provide excellent service	0.575	
	Factor 3		0.58
SO5	The training which I received prepared me to provide excellent service	−0.891	
SO9	My manager expects us to always follow procedures, even if it means giving less than excellent service	0.418	
SO7	At peak hours, we are so busy that we cannot provide excellent service	0.400	

Source: Data gained in the research.

Factor 2 can be interpreted as the dimension of *customer focus* (Dienhart, Gregoire, Downey, & Knight, 1992). Item SO2 has the strongest loading onto this factor (factor loading = 0.868). As expected, customer satisfaction is the primary representative of customer focus. Items SO1 and SO3, also have high loadings, i.e. the enjoyment during the contact with the hotel guests is of very important for the observed factor, as well as providing excellent service to guests. All these results show an expected relationship when it comes to customer focus, and we can easily conclude what the necessary measures are to raise this dimension in hotel facilities.

The training which hotel staff goes through in order to provide excellent service is highly negatively correlated with Factor 3 –*service under pressure* (factor loading = −0.891). This indicates a very disturbing fact that staff at Novi Sad's hotels claims that they do not have adequate training (in the work place) which will provide the best possible hotel service. Such an alarming indicator should be high priority of all hotel managers if they want to increase turnover and quality of hotel services. It is interesting to note that SO9 (factor loading = 0.418) has the highest positive correlation with service under pressure. This fact shows that hotel managers have expectations of adherence to service procedures even at the cost of providing low quality services. In conclusion, item SO7 indicates that the amount of work and customers affects quality of the service. This raises pressure on the employees which does not allow them to provide an adequate service.

According to Cohen's guidelines in interpreting strength of the relationship between variables (1988), there are significant positive correlations between job satisfaction and organisational support ($r = 0.52, p < 0.05$), on the one hand, job satisfaction and customer focus ($r = 0.36, p < 0.05$), on the other. Service under pressure does not correlate with job satisfaction ($r = 0.06$, $p > 0.05$) (see Table 6). These findings indicate that a higher level of organisational support and customer focus lead to greater satisfaction of the employed, while service under pressure does not influence job satisfaction.

Table 6. Correlation analysis – job satisfaction and service orientation.

	Job satisfaction
Organisational support	0.52
Customer focus	0.36
Service under pressure	0.06

Source: Data gained in the research.

The main problem of this article, considering gender differences in service orientation and job satisfaction, was analysed using t-test independent samples. Factors 1 and 2 of service orientation showed significant differences between men and women while Factor 3 of service orientation and job satisfaction did not show any gender differences (see Table 7).

Table 7. Gender differences in service orientation and job satisfaction (t-test).

	M factor loadings (scores) men	M factor loadings (scores) women	*t*-value	df	*p*
SO: Organisational support	0,246	−0,171	2,21	110	0,03
SO: Customer focus	−0,231	0.161	−2,07	110	0.04
SO: Service under pressure	0,077	−0.054	0,68	110	0.5
JS	20.717	19,969	1,07	110	0,3

Note: M – mean; t – value of t statistic; df – degrees of freedom for $p \leq 0.05$; p – level of statistical significance ($p < 0.05$).
Source: Data gained in the research.

The previous figure illustrates significant gender differences in organisational support and customer focus. According to the figure, we can notice that organisational support is more important to men, because men value more structured labour and service procedures than women ($t = 2.21$, $df = 110$, $p = 0.03$). On the other hand, customer focus is more relevant to a women, because it is more crucial for them to see satisfied clients who respect them ($t = 2.07$, $df = 110$, $p = 0.04$).

If we separate service orientation and job satisfaction by gender correlations, we can see that there are some specificities regarding relation of service under pressure and job satisfaction. Although these differences are not statistically significant, when providing service under pressure, in the same circumstances, women show a slight regression in job satisfaction while men show a slight progression (see Figure 3).

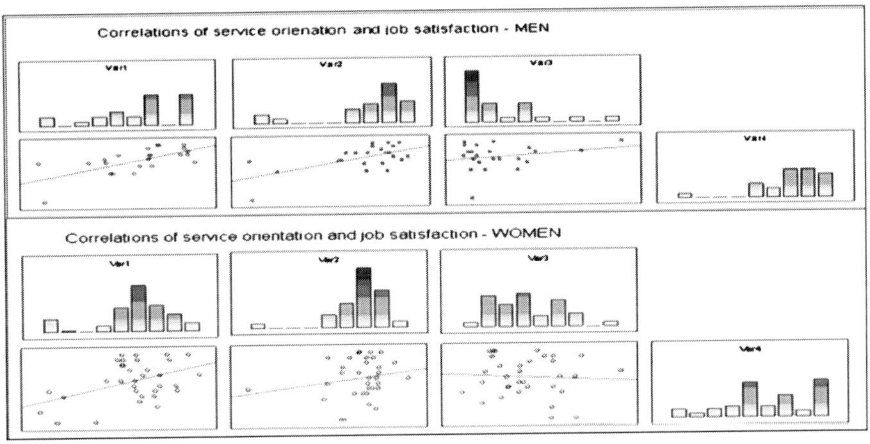

Note: Var1 = organisational support, Var2 = customer focus, Var3 = service under pressure, Var4 = job satisfaction.

Figure 3. Correlation of service orientation and job satisfaction – *men* and *women*.

Factor 1 corresponds to the component of *organisational support*, which seems to be more important to men than women, which is in accordance with our expected hypothesis. On the other hand, factor 2 in this case alludes to the component of*customer focus*, which is more relevant to female respondents of this research. This finding also confirms our initial assumptions. Factor 3, *service under pressure*, shows no gender difference, which disproves our starting hypothesis where we believed that men would be more resistant to providing service under pressure. Finally, level of *job satisfaction* was the same for both genders, which is in accordance with numerous previous research (Brief, Rose, & Aldag, 1977; De Vaus &

McAllister,1991; Mottazl, 1986; Smith, Smits, & Hoy, 1998; Tait, Padgett, & Baldwin, 1989; Ugorji, 1997; Witt & Nye, 1992). This also indicates our assumptions were erroneous, because we expected women to be more satisfied than men, like many authors proved previously (Clark, 1996, 1997; Hodson, 1989; Murray & Atkinson, 1981; Sloane & Williams, 1996).

7. RESEARCH DISCUSSION AND CONCLUSION

The goal of our study was to examine gender differences in service orientation and job satisfaction of 112 employees in 16 hotels in Novi Sad (Serbia). To determine these differences, we used the SOS, developed by Dienhart et al. (1992) and job satisfaction scale developed by Lytle (1994). We expected a three factor solution for the SOS and a unifactor solution for job satisfaction. More importantly, our assumptions were that there would be significant gender differences in service orientation and job satisfaction. Our research showed factor structures of the two scales we used. Specifically, we find gender differences in two of three factors, concerning service orientation, and no gender differences in job satisfaction.

7.1. Theoretical and practical implications

The factor solution for the SOS in our study is very similar to previous research which used this scale (Dienhart, Gregoire, Downey, & Knight, 1992; Kim, Leong, & Lee, 2005). Our Factor 1 resembles highly dimension *organisational support*. This dimension includes external motivation factors which come from the organisation itself. These factors are: high procedural structure, encouragement of the management, loyalty in the collective and good interpersonal relations. However, factors such as respect from the customers and risking quality service while under pressure also describe the Factor 1. This might be because manifested respect from customers usually leads to respect from the managers and other members of the staff which is a part of organisational support. Putting quality service as a priority instead of insisting on service, no matter what, also improves organisational support, in a long run. Quality service results in higher respect towards the product (hotel services) which leads to healthy organisational relations. Factor 2, which we identified as *customer focus*, involves motivation from within person. This means that employees value perceived satisfaction, kindness and respect from customers instead of support from the organisation. This dimension is very similar to the previous studies of Dienhart et al. (1992). Factor 3 shows similar structure as the dimension *service under pressure*. It includes working while service capacities are overloaded, blind following of the instructions and procedures and low level of previous training for providing hotel services. This indicates that this dimension includes all negative aspects of the service in hotels. Rigid following of procedures leads to potential conflicts with service consumers and other members of the staff which results in rising of tension and pressure on the workplace. Low level or complete lack of training for real situations in hotel practice makes employees feel inadequate and unprepared. Therefore, when they start working they feel pressure on a daily basis. So, all these aspects imply working under certain pressure which is in accordance with previous studies of Dienhart et al. (1992).

We also examined the correlation between three mentioned components of service orientation and job satisfaction. Findings replicated previous research of Kim et al. (2005). We concluded that job satisfaction has high correlation with *organisational support*, moderate correlation with *customer focus* and does not correlate with *service under pressure*. This can be explained by the fact that encouragement by supervisors, support from colleagues and friendly relationships, definitely affects the feeling of comfort and pleasantness at the workplace and potentially longer stay in the same working environment. All this might influence the professional development of each individual. Also, these circumstances contribute to staff enjoying while communicating with hotel guests, and they can stimulate staff to put added effort into the service process and this, in turn, improves their professional skills. It was noticed that job satisfaction does not significantly change when there is tension on the workplace, daily stress and pressure that is created from potentially unrealistic expectations from the business environment, mostly supervisors. However, it is observed that men showed statistically insignificant, but slight progress in job satisfaction when it comes to *service under pressure*, while women showed also statistically insignificant, but slight regress, which is not in accordance with previous research of Groves (1992). This can be explained by the fact that women are more sensitive to negative external effects, i.e. if they do not meet customer expectations they are more likely to become dissatisfied with their job and the workplace, than men. On the other hand, pressure at work stimulates men, since our results show that the increase of pressure leads to slight increase in job satisfaction, which can be described by a greater men' susceptibility to prove themselves. Also, men from this region have prominent competitive spirit (deep patriarchal roots), which is obviously reflected at the workplace. All these findings show factors that improve job satisfaction and they should be taken into account by hotel management.

When it comes to gender differences, we concluded that male hotel employees perceive more *organisational support*, which is not in accordance to previous findings of Kim, McCahon, and Miller (2003), who claimed that there is no significant gender impact on service orientation, or research of Groves (1992) who said that women perceive more support from the organisation. In our sample, the facts are that women also perceive organisational support, but not as much as men do. Female employees are more focused on users of hotel services and customer satisfaction is more important to them, as well as good relationship with guests. The job satisfaction of a female hotel employee will increase when she receives good comments and praise from customers and when she can see their satisfaction. Thus, interactions and feedback from customers are very important to them, which is in agreement with the finding of Groves (1992).

7.2. Benefits and contribution of the study

The research was of adequate sample size. From a total of 22 five-star hotel objects in Novi Sad we included 16 objects. The number of respondents was 112 hotel employees from eight different sectors of the organisational structure from the lowest level to executive management. If we take into account that the average number of employees in the city type hotels is 15 persons per object (Kosar, 2002), and if we multiply that number by the number of hotel objects, we get amount of about 330 employees in the hotels of Novi Sad. Therefore, we can claim that our research sample was sufficient to make generalisations on the entire population (over 1/3 of the total number of hotel employees in hotel enterprises of Novi Sad was included in the survey).

Our research is one of the first studies in Serbia that combines variables such as gender, service orientation and job satisfaction. Hotel staff in Novi Sad has long been scientifically neglected and this kind of research has great importance for understanding the complexity of service industries, such as hotel management. Also, gender differences were, until recently, almost unexplored in the hotel industry of Serbia. We believe that this article will contribute to the scientific understanding of obvious gender differences in two dimensions of service orientation of hotel employees: organisational support and customer focus and the importance of HR in hospitality.

7.3. Limitations of the study

Our suggestion for future research is to increase the sample size of hotel employees, so that gender differences in service orientation and job satisfaction might reach higher significance. This particularly goes for dimension service under pressure, whereby increasing the number of respondents, we might reach statistically significant gender effect.

There are some contradictions in interpreting items SO5 and SO4 in the SOS. For item SO5 ('The training which I received prepared me to provide excellent service.'), interpretation can go in two directions. The first direction is that respondents thought of their earlier formal education. Thus, respondents could mean that, during previous education (especially during high school and university), they did or did not have enough opportunities to adequately prepare for performing service procedures in a hotel. The second direction of interpretation of item SO5 may be that respondents thought that they did or did not have sufficient practical preparations and training to provide the best service, during their active work in the organisation. We chose the second direction of interpretation, but it is necessary to consider the contradiction of this claim and eventually specify this question in the future surveys. So, it is necessary to clarify: Is the omission made in the course of formal education or during training while working in the organisation? Or is the problem in both stages?

We would also like to highlight the item SO4 ('My manager encourages me to provide better service.'). The problem is in which way managers encourage their employees? Is it through financial incentives (material component), praise and showing respect (psychological component) or maybe through job promotion and prosperity (organisational component)? Finally, are there gender differences in this matter? Answers to these questions are important for the future research, i.e. it is important to specify the manner in which way hotel employees are encouraged by their supervisors.

Another problem that we would like to point out is highly local character of our research, so the future research should include other towns in Serbia and the surrounding countries (e.g. administrative centres, resort complexes, ski centres, spas, etc.). That way the entire study would have greater sample of respondents and data would be more relevant and comprehensive for chosen geographical area and perhaps for the entire country and Balkans. We opted for Novi Sad hotels because we wanted to explore the effect of service orientation and job satisfaction in city hotel types. This type of accommodation is characterised by the high frequency of guests (mainly business people) so work under pressure is everyday present due to the high demanding clientele. Also, in Serbia there are a small number of significant administrative centres, and Novi Sad represents one of the biggest.

Although, there are limitations within this study which demand further research, certain benefits arise from our results. Job satisfaction and service orientation are two very important constructs for hotel industry. There are gender specific differences in customer focus and organisational support. However, job satisfaction is the same for men and women. So, it is of great importance for hotel management to have in mind that man and women respond and behave differently depending on the stimuli in the work environment and their satisfaction varies because of it.

ACKNOWLEDGMENT

The research was supported by Ministry of Education, Science and Technological Development, Republic of Serbia (Grants 47007 and 176020).

REFERENCES

1. Arnett, D. B., Laverie, D. A., & McLane, C. (2002). Using job satisfaction and pride as internal-marketing tools. *Cornell Hotel and Restaurant Administration Quarterly, 43,* 87–96. doi:10.1016/S0010-8804(02)80035-010.1016/S0010-8804(02)80035-0

2. Arvey, R. D., Bouchard, T. J., Segal, N. L., & Abraham, L. M. (1989). Job satisfaction: Environmental and genetic components.*Journal of Applied Psychology, 74,* 187–192. doi:10.1037/0021-9010.74.2.18710.1037/0021-9010.74.2.187

3. Ashforth, B. E., & Humphrey, R. H. (1993). Emotional labor in service roles: The influence of identity. *Academy of Management Review, 18,* 88–115. doi:10.5465/AMR.1993.3997508

4. Aydin, A., Uysal, S., & Sarier, Y. (2012). The effect of gender on job satisfaction of teachers: A meta-analysis study. *Procedia Social and Behavioral Sciences, 46,* 356–362. doi:10.1016/j.sbspro.2012.05.12210.1016/j.sbspro.2012.05.122

5. Aykac, A. (2006). Linking scale of operation and labor transformation on tourism: Comparative analysis on three Turkish cases. *Anatolia: An International Journal of Tourism and Hospitality Research, 17,* 257–278. doi:10.1080/13032917.2006.9687189

6. Babin, B. J., & Boles, J. S. (1996). The effects of perceived co-worker involvement and supervisor support on service provider role stress, performance and job satisfaction. *Journal of Retailing, 72,* 57–75. doi:10.1016/S0022-4359(96)90005-610.1016/S0022-4359(96)90005-6

7. Bandura, A. (1982). The psychology of chance encounters and life paths. *American Psychologist, 37,* 747–755. doi:10.1037/0003-066X.37.7.74710.1037/0003-066X.37.7.747

8. Birnbaum, D., & Somers, M.J. (1993). Fitting job performance into turnover model: An examination of the form of the job performance – Turnover relationship and path model. *Journal of Management, 19*(1), 1–11.

9. Blešić, I. Ivkov-Džigurski, A., Dragin, A., Ivanović, L., & Pantelić, M. (2011). Application of gap model in the researches of hotel services quality. *Turizam, 15*, 40–52. YU ISSN 1450-6661

10. Blešić, I., Popov-Raljić, J., & Romelić, J. (2007). Kvalitet usluga u novosadskim hotelijerskim objektima [Service Quality in Novi Sad's Hotel Objects]. *Turizam, 11*, 210–213. YU ISSN 1450-6661

11. Bolton, S., & Boyd, C. (2003). Trolley dolly or skilled emotion manager? Moving on from hochschild's managed heart. *Work, Employment and Society, 17*, 289–308.

12. Brief, A. P., Rose, G. L., & Aldag, R. J. (1977). Sex differences in preferences for job attributes revisited. *Journal of Applied Psychology, 62*, 645–646.

13. Čačić, K. (1998). *Poslovanje preduzeća u turizmu* [Enterprises' business in tourism]. Belgrade: Faculty of Economy, University of Belgrade.

14. Campos-Soria, J. A., Marchante-Mera, A., & Ropero-García, M. A. (2011). Patterns of occupational segregation by gender in the hospitality industry. *International Journal of Hospitality Management, 30*, 91–102.

15. Carmeli, A. (2003). The relationship between emotional intelligence and work attitudes, behavior and outcomes. *Journal of Managerial Psychology, 18*, 788–813.

16. Caruso, D. R., Mayer, J. D., & Salovey, P. (2002). Relation of an ability measure of emotional intelligence to personality. *Journal of Personality Assessment, 79*, 306–320.

17. Cave, P., & Kilic, S. (2010). The role of women in tourism employment with special reference to Antalya, Turkey. *Journal of Hospitality Management and Marketing, 19*, 280–292.

18. Čerović, S. (2002). *Strategijski menadžment turističke privrede Srbije* [Strategic management of tourism in Serbia]. Belgrade: Želind.

19. Cheng, C. C., & Krumwiede, D. (2012). The role of service innovation in the market orientation-new service performance linkage. *Technovation, 32*, 487–497.

20. Chiang, F., & Birtch, T. A. (2011). The impact of reward climate on service quality orientation and employee attitudes. *International Journal of Hospitality Management, 30*, 3–9.

21. Clark, A. E. (1996). Job satisfaction in Britain. *Journal of Industrial Relations, 34*, 189–217.

22. Clark, A. E. (1997). Job satisfaction and gender: Why are women so happy at work? *Labor Economics, 4*, 341–372.

23. Cohen, J. (1988). *Statistical power analysis for the behavioral sciences* (2nd ed.). Hillsdale, NJ: Lawrence Erlbaum Associates, Publishers.

24. De Vaus, D., & McAllister, I. (1991). Gender and work orientation: Values and satisfaction in Western Europe. *Work and Occupation, 18*, 72–93.

25. Deutsch, F. M. (1990). Status, sex, and smiling: The effect of role on smiling in men and women. *Personality and Social Psychology Bulletin, 16*, 531–540.

26. Dienhart, J. R., Gregoire, M.B., Downey, R. G., & Knight, P. K. (1992). Service orientation of restaurant employees.*International Journal of Hospitality Management, 11*, 331–346.

27. Eisenberger, R., Huntington, R., Hutchison, S., & Sowa, D. (1986). Perceived organizational support. *Journal of Applied Psychology, 75*, 51–59.

28. Galičić, V., & Ivanović, S. (2008). *Menadžment zadovoljstva gosta* [Guest Pleasure Management]. Opatija: Faculty of Management in Tourim and Hospitality.

29. García-Pozo, A., Campos-Soria, J. A., Sánchez-Ollero, J. L., & Marchante-Lara, M. (2012). The regional wage gap in the Spanish hospitality sector based on a gender perspective. *International Journal of Hospitality Management, 31*, 266–275.

30. George, J. M. (2000). Emotions and leadership: The role of emotional intelligence. *Human Relations, 53*, 1027–1055.

31. Groves, J. (1992). *Perceived service orientation of restaurant employees* [Unpublished doctoral dissertation]. Manhattan, KS: Kansas State University.

32. Hartline, M. D., & Ferrell, O. C. (1996). The management of customer-contact service employees: An empirical investigation. *Journal of Marketing, 60*, 52–70.

33. Hochschild, A. (1983). *The managed heart: Commercialization of human feeling*. Berkeley: University of California Press.

34. Hodson, R. (1989). Gender differences in job satisfaction: Why aren't women more dissatisfied? *Sociological Quarterly, 30*, 385–399.

35. Hoffman, L. W. (1972). Early childhood experiences and women's achievement motives. *Journal of Social Issues, 28*, 129–155.

36. Hoque, K. (2000). *Human resource management in the hotel industry. Strategy, innovation and performance.* London: Routledge.

37. Humphreys, M. A., & Williams, M. R. (1996). Exploring the relative effects of salesperson interpersonal process attributes and technical product attributes on customer satisfaction. *Journal of Personal Selling and Sales Management, 16*, 47–57.

38. Hunt, S. D., & Chonko, L. B. (1984). Marketing and Machiavellianism. *Journal of Marketing, 48*, 30–42.

39. Ishitani, T. T. (2010). Exploring the effects of congruence and Holland's personality codes on job satisfaction: An application of hierarchical linear modeling techniques. *Journal of Vocational Behavior, 76*, 16–24.

40. Iverson, R. D. (2000). An event history analysis of employee turnover: The case of hospital employees in Australia. *Human Resource Management Review, 9*, 397–418.

41. James, N. (1989). Emotional labour: Skill and work in the social regulation of feelings. *Sociological Review, 37*, 15–42. doi:10.1111/j.1467-954X.1989.tb00019.x10.1111/sore.1989.37.issue-1

42. Jaworski, B. J., & Kohli, A. K. (1993). Market orientation: Antecedents and consequences. *Journal of Marketing, 57*, 53–70.

43. Jerome, L., & Kleiner, B. H. (1995). Employee morale and its impact on service: What companies do to create a positive service experience. *Managing Service Quality, 5*, 21–22.

44. Kim, H. J. (2008). Hotel service provider's emotional labour: The antecedents and effects on Burnout. *International Journal of Hospitality Management, 27*, 151–161.

45. Kim, W. G., Leonga, J. K., & Lee, Y. K. (2005). Effect of service orientation on job satisfaction, organizational commitment, and intention of leaving in a casual dining chain restaurant. *International Journal of Hospitality Management, 24*, 171–193.

46. Kim, H. J., McCahon, C., & Miller, J. (2003). Service orientation for contact employees in Korean casual-dining restaurants.*International Journal of Hospitality Management, 22*, 67–83.

47. Kim, B. P., Murrmann, S. K., & Lee, G. (2009). Moderating effects of gender and organizational level between role stress and job satisfaction among hotel employees. *International Journal of Hospitality Management, 28*, 612–619.

48. Kosar, L. (2002). *Hotelijerstvo* [Hotel Industry]. Belgrade: College of Hotel Management.

49. Kottke, J. L., & Sharafinski, C. E. (1988). Measuring perceived supervisory and organizational support. *Educational and Psychological Measurement, 48*, 1075–1079.

50. Kruml, S. M., & Geddes, D. (2000). Exploring the dimensions of emotional labor: The heart of Hochschild's work.*Management Communication Quarterly, 14*, 8–49.

51. LaFrance, M., & Banaji, M. (1992). Toward a reconsideration of the gender-emotion relationship. *Review of Personality and Social Psychology, 14*, 178–197.

52. Lahey, M. A. (1984).*Job Security: Its meaning and measure (Unpublished Doctoral Dissertation)*. Kansas City, KS: Kansas State University.

53. Lawler, E. E., & Suttle, L. (1973). Expectancy theory and job behavior. *Organizational Behavior and Human Performance, 9*, 482–503.

54. Lee, Y. K., Park, D. H., & Yoo, D. K. (1999). The structural relationships between service orientation, mediators, and business performance in Korea hotel firms. *Asia Pacific Journal of Tourism Research, 4*, 59–70.

55. Leidner, R. (1993). *Fast food, fast talk: Service work and the routinization of everyday life*. Berkeley: University of California.

56. Loscocco, K. (1989). The interplay of personal and job characteristics in determining work commitment. *Social Science Research, 18*, 70–79.

57. Lucas, G. H., Parasuraman, A., Davis, R. A., & Enis, B. M. (1987). An empirical study of salesforce turnover. *Journal of Marketing, 51*, 34–59.

58. Lytle, R. S. (1994). *Service orientation, market orientation, and performance: An organizational culture perspective* (Unpublished Doctoral Dissertation). Arizona: Arizona State University.

59. Lytle, R. S., Hom, P. W., & Mokwa, M. P. (1998). SERV*OR: A managerial measure of organizational service orientation.*Journal of Retailing, 74*, 447–454. doi:10.1016/S0022-4359(99)80104-3

60. Malka, A., & Chatman, J. A. (2003). Intrinsic and extrinsic work orientations as moderators of the effect of annual income on subjective well-being: A longitudinal analysis. *Personality and Social Psychology Bulletin, 29*, 737–746.

61. Mobley, W. H. (1977). Intermediate linkages in the relationship between job satisfaction and employee turnover. *Journal of Applied Psychology, 62*, 237–240.

62. Mora, T., & Ferrer-i-Carbonell, A. (2009). The job satisfaction gender gap among young recent university graduates: Evidence from Catalonia. *Journal of Socio-Economics, 38*, 581–589.

63. Morris, J., & Feldman, D. (1997). Managing emotions in the workplace. *Journal of Managerial Issues, 9*, 257–274. ISSN: 1045-3695

64. Mottazl, C. (1986). Gender differences in work satisfaction, work related rewards and values, and the determinants of work satisfaction. *Human Relations, 39*, 359–378.

65. Murray, M., & Atkinson, T. (1981). Gender differences in correlates of job satisfaction. *Canadian Journal of Behavioral Sciences, 13*, 44–52.

66. Nedeljković, M. (2008). *Istraživanje motivacije i zadovoljstva poslom u organizaciji u uslovima tranzicije* [The study of motivation and job satisfaction in the organization in transition period]. (Unpublished doctoral dissertation), Faculty of Technical Sciences, University of Novi Sad, Serbia.

67. Neil, C. C., & Snizek, W. E. (1987). Work values, job characteristics, and gender. *Sociological Perspectives, 30*, 245–265.

68. Nikolić, M. (2006). Uloga menadžmenta hotela u preispitivanju procesa rada [The role of hotel management in review of work process]. *Turizam, 10*, 176–178. YU ISSN 1450-6661

69. O'Connor, S. J., Trinh, H. Q., & Shewchuk, R. M. (2000). *Determinants of service orientation among medical students.* Oxford, OH: Miami University Farmer School of Business.

70. Odom, R. Y., Boxx, W. R., & Dunn, M. G. (1990). Organizational culture, commitment, satisfaction and cohesion. *Public Productivity and Management Review, 14*, 157–168.

71. Petrović, M., & Marković, J. (2011). *Correlation between Service Orientation and Job Satisfaction among the Employees in Novi Sad's Hotels*, Paper presented at the meeting of International Conference "Contemporary Trends in Tourism and Hospitality", Novi Sad, Serbia – collection of papers, 374–386.

72. Pinar, M., McCuddy, M. K., Birkan, I., & Kozak, M. (2011). Gender diversity in the hospitality industry: an empirical study in Turkey. *International Journal of Hospitality Management, 30*, 73–81.

73. Purcell, K. (1996). The relationship between career and job opportunities: Women's employment in the hospitality industry as a microcosm of women's employment. *Women in Management Review, 11,* 17–24. ISSN: 0964-9425

74. Rafaeli, A. (1989). When clerks meet customers: A test of variables related to emotional expression on the job. *Journal of Applied Psychology, 74,* 389–393.

75. Rhoades, L., & Eisenberger, R. (2002). Perceived organizational support: A review of the literature. *Journal of Applied Psychology, 87,* 698–714.

76. Schneider, B., & Bowen, D. E. (1985). Employee and customer perceptions of service in banks: Replication and extension. *Journal of Applied Psychology, 70,* 423–433. ISSN: 0001-839210.1037/0021-9010.70.3.423

77. Shaw, J. D. (1999). Job satisfaction and turnover intentions: The moderating role of positive affect. *The Journal of Social Psychology, 139,* 242–244.

78. Sloane, P., & Williams, H. (1996). Are 'overpaid' workers really unhappy? A test of the theory of cognitive dissonance. *Labour, 10,* 3–15. ISSN:

79. Smith, P. L., Smits, S. J., & Hoy, F. (1998). Employee work attitudes: The subtle influence of gender. *Human Relations, 51,* 649–666. **80.** Sousa-Poza, A., & Sousa-Poza, A. A. (2007). The effect of job satisfaction on labor turnover by gender: An analysis for Switzerland. *Journal of Socio-Economics, 36,* 895–913.

80. Sujan, H. (1986). Smarter versus harder: An exploratory attributional analysis of salespeople's motivation. *Journal of Marketing Research, 23,* 41–49. ISSN: 0022-243710.2307/3151775

81. Tait, M., Padgett, M., & Baldwin, T. (1989). Job and life satisfaction: A reevaluation of the strength of the relationship and gender effects as a function of the date of the study. *Journal of Applied Psychology, 74,* 502–507.

82. Teas, K. R. (1983). Supervisory behavior, role stress, and the job satisfaction of industrial salespeople. *Journal of Marketing Research, 20,* 84–91.

83. The Statistical Office of the Republic of Serbia (http://webrzs.stat.gov.rs/WebSite/)

84. Ugorji, U. O. (1997). Career-impeding supervisory behaviors: Perceptions of African American and European American professionals. *Public Administration Review, 57,* 250–255.

85. Wharton, A. S. (1999). The psychosocial consequences of emotional labor. *Annals of the American Academy of Political and Social Science, 561*, 158–176.

86. Wharton, A. S., & Erickson, R. J. (1993). Managing emotions on the job and at home: Understanding the consequences of multiple emotional roles. *Academy of Management Review, 18*, 457–486.

87. Witt, L. A., & Nye, L. G. (1992). Gender and the relationship between perceived fairness of pay or promotion and job satisfaction. *Journal of Applied Psychology, 77*, 910–917.

Chapter 5

On Becoming a Strategic Partner: The Role of Human Resources in Gaining Competitive Advantage

Jay B. Barney, Patrick M. Wright

Faculty of Management and Human Resources The Ohio State University Columbus, School of Industrial and Labor Relations Cornell University Ithaca, NY 14853-3901

ABSTRACT

Although managers cite human resources as a firm's most important asset, many organizational decisions do not reflect this belief. This paper uses the VRIO (value, rareness, imitability, and organization) framework to examine the role that the Human Resource (HR) function plays in developing a sustainable competitive advantage. We discuss why some popularly cited sources of sustainable competitive advantage are not, and what aspects of a firm's human resources can provide a source of sustainable competitive advantage. We also examine the role of the HR executive as a strategic partner in developing and maintaining competitive advantage within the firm.

Keywords: manager, human resource, firm, value, organization, function, HR, advantage

Human resource researchers and managers have long maintained that the human resource function plays an important role in firm performance. In fact, most corporate annual reports boldly state that the firm's people are its most important asset. However, despite these widely-held beliefs and all-too-frequent statements, many organizational decisions suggest a relative low priority on both the human resources of the firm and the

Human Resource (HR) department. For example, when organizations require cost cutting, they look first to investments in the firm's people such as training, wages, and headcounts.

In addition, even when top managers value the firm's people, they may not value the HR department. For example, when asked how the founder and CEO of one of the most successful high technology companies in the world viewed the importance of human resources, the Director of Strategic Leadership Development replied, "Which do you mean? If you mean the Human Resource function, or what we call 'big HR,' then he doesn't have much value for them at all. If you mean the people of the company, or what we call 'little hr,' then he places an extremely high value on them."

If top managers publicly espouse their commitment to the firm's human resources, and the firm's HR function has substantial responsibility for managing this valuable firm resource, then why do many organizational decisions not evidence this stated commitment to people or a respect for the HR function? We believe that the fault lies, in part, with the fact that few HR executives can explain, in economic terms, how a firm's people can provide sustainable competitive advantage, and the role that the HR function plays in this process. Furthermore, due to this lack of understanding, many HR executives fail to direct the HR activities toward developing characteristics of the firm's human resources that can be a source of sustainable competitive advantage.

In this paper we will examine the economics underlying the role of human resources in a firm's competitive advantage. We will discuss the Resource-Based View of the Firm (Barney, 1991; Wernerfelt, 1984) and examine the VRIO framework (Barney, 1995; 1996) for analyzing sources of competitive advantage and sustainable competitive advantage. Finally, we will discuss the implications of the VRIO framework for the Human Resource executive who seeks to transform the HR function in his/her organization into a source of competitive advantage. The analysis will provide executives in the HR function with the tools necessary to analyze how they can manage the function to develop a firm's people as a source of sustainable competitive advantage.

THE RESOURCE BASED VIEW OF COMPETITIVE ADVANTAGE

Academics and managers have both sought to understand the factors that determine the profitability of firms for many years. The resource-based view of organizations provides an economic foundation for examining the role of HR in firm competitive advantage. This view focuses on firm resources that can be sources of competitive advantage within the industry (Barney, 1995). Three basic types of resources can provide competitive advantage (Barney, 1991). Physical capital resources include such things as the firm's plant, equipment, and finances. Organizational capital resources consist of such things as the firm's structure, planning, controlling, coordinating and HR systems. Finally, human capital resources include such things as the skills, judgment, and intelligence of the firm's employees.

Because of its recognition of the potential for human assets of organizations to provide competitive advantage, many academic authors have applied the resource based view to understanding the role of HR in organizations. For example, Wright, McMahan, & McWilliams (1994) used the resource based framework to examine how a firm's human resources can be a source of sustainable competitive advantage. Lado and Wilson, (1994) explored the potential for HR practices to be a source of competitive advantage. Snell, Youndt, and Wright (1996) attempted to integrate the resource based view with the concept of organizational learning. Additional applications of the resource based view to human resources are discussed in Jackson and Schuler (1995).

For the sake of our discussion of the role of HR in competitive advantage, similar to Wright et al. (1994), we will focus on the characteristics of a firm's human resources, including all of the knowledge, experience, skill, and commitment of a firm's employees, and their relationships with each other and those outside the firm. While many implicitly believe that these characteristics can provide a firm with a source of competitive advantage, often it is difficult to specify which characteristics do so. In addition we consider HR practices as including all of the programs, policies, and practices that firms use to manage their human resources.

With this in mind, we propose that to identify the value of a firm's human resources to organizations as well as the proper role of the HR function in managing the firm's human resources to achieve such an advantage, one needs to ask four questions. These questions include the questions of Value, Rareness, Imitability, and Organization, or what is referred to as the VRIO framework (Barney, 1995).

THE VRIO FRAMEWORK

The Question of Value

Firms create value through either decreasing product/service costs or differentiating the product/service in a way that allows the firm to charge a premium price. Thus, the ultimate goal of any HR executive is to create value through the human resource function. The first question that an HR executive must address is "How can the HR function aid in either decreasing costs or increasing revenues?"

Alcon Laboratories exemplifies the role of HR practices in directly decreasing costs. Trying to hold down the cost of health insurance, Alcon sought to encourage employees to take part in the less expensive Preferred Provider Organization (PPO) rather than the traditional fee for service type plans. VP of HR Jack Walters noticed that many doctors who were part of the PPO were not the doctors currently used by employees. Thus, in negotiating with MetLife, he asked them to identify the doctors Alcon employees were using and recruit those doctors into the PPO. MetLife was able to bring most of those doctors into their PPO, and as a result, Alcon's health insurance costs have increased at less than half of the industry average.

Increasing revenues, on the other hand, is a more foreign goal to HR managers, but one in which they can play an important role. For example, Federal Express (now FedEx) illustrates the value created by human resources. Federal Express managers stress that they are a "people-first" organization. The corporate philosophy statement sums up their view of the source of competitive advantage: "People-Service-Profit." Fred Smith, founder and CEO of the firm, says "We discovered a long time ago that customer satisfaction really begins with employee satisfaction," (Waterman, 1994). In other words, the FedEx philosophy is that people are the primary link in the value chain, and thus, value is created by focusing on employees first.

How is this operationalized to create value? This emphasis on employee satisfaction is illustrated by their annual attitude survey. Most organizations administer attitude surveys from time to time, and occasionally use the information gleaned from the surveys to address the most glaring organizational problems. However, at FedEx, the attitude survey forms part of the annual managerial evaluation and reward process. The survey addresses the atmosphere of an individual's immediate work group, the immediate manager, the managers at levels higher in the organization, and the company's atmosphere in general. Scores on the items covering the work group and the immediate manager forms "the leadership index."

This index is used in two ways. If an individual manager receives low scores on the index from the employees reporting to him or her, that manager faces a year long probation. During that time the manager is expected to improve the scores to an acceptable level or face some type of punitive action. Second, each year a goal is set for the company's score on the leadership index. If the goal is not met, the top 300 managers in the firm do not receive any bonus, which usually is about 40% of base salary. By linking rewards and punishment to employee satisfaction levels, the firm ensures that employees are treated right. When they are treated right, they treat customers right, and create value.

FedEx's philosophy has gained an increasing base of empirical support. For example, Schneider and Bowen (1985) hypothesized that HR practices would be related to employee attitudes which would be consequently related to customer satisfaction. They found significant relationships between HR practices and customer reports of the quality of service they received in a sample of banks. Schlesinger and Zornitsky (1991) found that job satisfaction predicted employee's perceptions of service quality as well as the discrepancy between employee and customer perceptions of quality. Ulrich, Halbrook, Meder, Stuchlik, and Thorpe (1991) found significant relationships between the tenure of employees and customer satisfaction. Tornow and Wiley (1991) found that employee attitudes such as job satisfaction were related to measures of organizational performance. Finally, most recently, Schmit and Allscheid (1995) found that employees' climate perceptions of management, supervisor, monetary, and service support were related to employee affect. Affect was related to service intentions, which was related to customer service. Thus, empirical research supports the notions that employee satisfaction is linked to service quality, and that HR practices are important determinants of employee satisfaction.

Finally, some HR practices can impact both costs and revenues. Continental Airlines has recently experienced a tremendous turnaround in which the HR function played a vital role. One of the frequently cited HR practices responsible for this turnaround was the on-time bonus, an incentive system where each employee was paid a bonus of $65 for every month the airline was at the top of the industry in on-time performance (Boisseau, 10/22/95). While this may seem like it comes straight from any introductory textbook (Barlow, 4/7/96), it origin was not nearly so simple. In early 1995, after years of pay cuts or no pay raises, top management discovered that it again would be unable to give pay raises to employees. HR executives recognized that taking that message to the employees at a critical phase of the turnaround would destroy morale and greatly impede the cultural shift under way. Thus, HR executives along with line executives came up with the idea of the on-time bonus.

This bonus has resulted in Continental moving from last to first in the industry in on-time performance, and consequently has both decreased costs and increased revenues. On the cost side, last year it paid out $51 million in bonuses, but saved $75 million in lower passenger accommodation costs such as money for meals and hotel rooms associate with missed connections. On the revenue side, the bonus has been instrumental in restoring employee morale and thus, increasing customer satisfaction. In addition, because on-time performance is an important criterion for the higher revenue business traveler, this bonus has a strong impact on the firm's revenues as they have increased their share of the business traveler market.

The Question of Rareness

The value of a firm's human resources is a necessary but not sufficient criteria for competitive advantage. If the same characteristic of human resources is found in many competing firms, then that characteristic cannot be a source of competitive advantage for any one of them. Valuable but common characteristics of human resources provide only competitive parity, ensuring that a firm is not at a substantial competitive disadvantage because it does not possess that characteristic. Thus, an HR executive must examine how to develop and exploit rare characteristics of the firm's human resources to gain competitive advantage.

For example, most firms view the labor pool for particular jobs as relatively homogeneous. However, within any labor pool, difference exist across individuals in terms of their job related skills and abilities. Thus, if the assumption exists across firms that the labor pool is homogeneous, there would be tremendous potential to exploit the rare characteristics of those employees for competitive advantage (Wright et al., 1994).

For example, Nordstrom's exists in the highly competitive retailing industry. This industry is usually characterized as having relatively low skill requirements and high turnover for sales clerks. However, Nordstrom's has attempted to focus on individual salespersons as the key to its competitive advantage. They invest in attracting and retaining young, college-educated sales clerks who desire a career in retailing. They provide a highly incentive based compensation system that allows Nordstrom salespersons to make as much as twice the industry average in pay. The Nordstrom culture encourages sales clerks to make heroic efforts to attend to customers' needs, even to the point of changing a customer's flat tire in the parking lot. The recruiting process, compensation practices and culture at Nordstrom's have helped the organization to maintain the highest sales per square foot of any retailer in the nation. Thus, Nordstrom's has taken what is considered to be a relatively homogeneous labor pool and exploited the rare characteristics of their employees to gain a competitive advantage.

The Question of Imitability

Valuable and rare characteristics of a firm's human resources can provide above normal profits for the firm in the short term. However, if other firms can imitate these characteristics, then over time the characteristics will provide no more than competitive parity. Thus, the HR executive must attempt to develop and nurture characteristics of the firm's human resources that cannot easily be imitated by competitors. This points to focusing on the importance of socially complex phenomena such as an organization's unique history or culture in providing competitive advantage.

Every firm has a unique history that defines the present situation. This history often provides a foundation for a competitive advantage which other firms will find impossible to imitate. For example, in a recent conversation, a high-level executive at one of DuPont's competitors bemoaned the fact that no matter what his firm did (including purchasing DuPont's safety training programs), they were simply unable to match DuPont's safety record. When asked why, his response was simply, "When a firm starts out by making dynamite, something happens that just instills in employees' minds the importance of safety." Thus, DuPont's superior safety performance stems at least partly from its unique history that competitors would find impossible to imitate.

Southwest Airlines exemplifies the role that socially complex phenomena such as culture plays in competitive advantage. According to the company's top management, the firm's success can be attributed to the "personality" of the company; a culture of fun and trust that provides employees with both the desire and the discretion to do whatever it takes to meet the customers' needs. The "Fun" airline uses an extensive selection process for hiring flight attendants who will project the Fun image of the airline. Applicants must go through a casting call type exercise where they are interviewed by a panel that includes current flight attendants, managers, and customers. The applicants tell stories such as their most embarrassing experience in front of the panel and other applicants. Those who make it through the panel interview are then examined against a psychological profile that distinguished outstanding past flight attendants from those who were mediocre or worse.

In addition to the extensive selection process, employees are empowered to create an entertaining traveling environment by a strong organizational culture that values customer satisfaction. Says Herb Kelleher, CEO, "We tell our people that we value inconsistency. By that I mean that we're going to carry 20 million passengers this year and that I can't foresee all of the situations that will arise at the stations across our system. So what we tell our people is, 'Hey, we can't anticipate all of these things, you handle them the best way possible. You make a judgment and use your discretion; we trust you'll do the right thing. If we think you've done something erroneous, we'll let you know--without criticism, without backbiting.'" (Quick, 1992). This extensive selection process and the strong organizational culture contribute to the differentiated service that has made Southwest Airlines the most financially successful airline over the past 20 years and continually being among the best in the industry for having the fewest customer complaints.

Seeing this financial success, competitors such as Continental Airlines (Continental Lite) and United Airlines (United Express) attempted to compete with Southwest Airlines by providing low cost service to a number of destinations. However, Continental Lite ceased operations within a year, and United, while having survived, is still losing to Southwest in most markets where they compete.

Before they even entered the market, the VRIO framework could have accurately predicted the success potential for such a competitive response. Kelleher believes that Southwest's superior performance has happened because its culture simply cannot be imitated. Kelleher summarizes the resource based view of competitive advantage when he states, "Maybe someone could equal the cost...possibly they could. And maybe someone could equal the quality of service that goes along with that and constitutes great value...possibly they could. But the one thing they would find it impossible to equal very easily is the spirit of our people and the attitude they manifest toward our customers." In other words, the human resources of Southwest Airlines serves as a source of sustained competitive advantage because they create value, are rare, and are virtually impossible to imitate.

The Question of Organization

Finally, in order for any characteristic of a firm's human resources to provide a source of sustained competitive advantage, the firm must be organized to exploit the resource. Organization requires having in place the systems and practices that allow human resources characteristics to bear the fruit of their potential advantages.

For example, both General Motors and Ford have historically recruited assembly line workers from the same basic labor market. There is little evidence that the skill levels of Ford's workers are significantly higher than those of General Motors. However, Ford has been more successful at developing a cooperative, team-based culture than General Motors. Both automakers set out to develop employee involvement programs during the late 1970's and early 1980's. However, Ford more successfully changed the culture and HR systems to allow for, and even value employee participation in decision making relative to GM. Ford's culture and HR systems allow for employees to participate in decision making and to utilize cognitive skills that the GM systems have been less able to exploit (Templin, 12/15/1992). In addition, as Ford moves toward hiring even more highly skilled employees through an extensive assessment process, its participative system will leave it poised to increase its relative advantages over GM (Templin, 3/10/1994).

The question of organization focuses attention on systems, as opposed to single HR practices. Recent work on HR practices and firm performance seems to indicate that HR practices are maximally effective when they exist as a coherent system. Wright & Snell (1992) argued that Strategic Human Resource Management required coordinated HR activities across the various subfunctions. Similarly, Wright and McMahan's (1992) definition of SHRM called for "horizontal integration" of the various HR practices rather than viewing each in isolation. Lado and Wilson, (1994) hypothesized that the more complex the HR system, the more likely it would be to serve as a source of sustainable competitive advantage. MacDuffie (1995), in a study of automobile manufacturing firms found that performance was maximized when "bundles" of HR practices were linked with participative work systems and flexible production systems. Wright, McCormick, Sherman, & McMahan (1996) found that HR practices such as selection, appraisal and compensation, were unrelated to the financial performance of petro-chemical refineries alone, but that they were strongly positively related to performance among refineries that had highly participative work systems These research studies seem to indicate a need for HR functions to pay attention to the system of HR practices, rather than to focus on each in isolation.

However, both quantitative and qualitative data gathered from an ongoing research study conducted by the second author indicates that very few companies are spending much time and attention on coordinating each of the various HR subfunctions (e.g., staffing, compensation, training, etc.) with one another. Of 13 firms in the study, only 2 have actively attended to achieving integration among the compensation, selection, training, and appraisal systems and processes. Thus, it appears that firms that do make such efforts have at least temporary advantages over their competitors.

Thus, these examples illustrate how the VRIO framework can be used to analyze the ways in which a variety of firms have attempted to develop their human resources as a source of sustainable competitive advantage. Figure 1 illustrates how to use the VRIO framework to analyze the potential for firm resources to be sources of competitive disadvantage, competitive parity, competitive advantage, and sustained competitive advantage. According to this framework, aspects of human resources that do not provide value can only be a source of competitive disadvantage. These resources or activities are ones that HR executives should be discarding from the HR function. Aspects of human resources that provide value, but are not rare are sources of competitive parity. These resources are not to be dismissed as useless. Not to have them is a source of competitive disadvantage, but, because other firms possess them, they cannot provide and advantage in the competitive arena. Temporary competitive advantage stems from resources that provide value and are rare, but are easily imitated. However, if these resources do serve as a source of competitive advantage, then other firms will soon imitate them, resulting in competitive parity. Finally, aspects of human resources that are valuable, rare, and uneasily imitated, can be a source of sustained competitive advantage, but only if the firm is organized to capitalize on these resources.

Thus, clearly the HR function, through either directly controlling or strongly influencing the characteristics of human resources in organizations plays an important role in developing and maintaining a firm's competitive advantage. However, simply making the case that HR can influence a firm's performance is only part of the story. In order for HR to truly develop and maintain sources of competitive advantage, HR executives need to focus attention and activities toward those aspects of the firm's resources that will provide such an advantage. In the next section we will discuss the implications of the VRIO framework sources of sustained competitive advantage that might be influenced by leaders of the HR function in organizations.

Is a Resource...

Valuable?	Rare?	Difficult to Imitate?	Supported by Organization?	Competitive Implications	Performance
No	----	----		Competitive Disadvantage	Below Normal
Yes	No	----		Competitive Parity	Normal
Yes	Yes	No		Temporary Competitive Advantage	Above Normal
Yes	Yes	Yes		Sustained Competitive Advantage	Above Normal

Figure 1. The VRIO Framework

Implications for Competitive Advantage

The VRIO framework presents implications for what types of resources both can and cannot be sources of sustainable competitive advantage.

Sustainable Competitive Advantage Stems from Firm Specific more than General Skills.

Human Capital theory (Flamholtz & Lacey, 1981) distinguished between general skills and firm specific skills of human resources. General skills are skills possessed by individuals that provide value to a firm and are transferable across a variety of firms. For example, all competitor firms have the potential to accrue equal value from acquiring employees with knowledge of general management, the ability to apply financial ratios, or general cognitive ability. Specific skills, on the other hand, provide value only to a particular firm, and are of no value to competing firms. For example, the knowledge of how to use a particular technology used only by one firm, or knowledge of a firm's policies and procedures provide value to that firm, but usually would not be valuable to other firms.

Because general skills provide equal value to all firms, one would expect that, given even moderately efficient labor markets, these would not be a source of competitive advantage for any one organization. Thus, to seek to gain sustained competitive advantage through general skills would be futile. On the other hand, this does not imply that these skills are not important for two reasons. First, general skills are necessary for maintaining competitive parity. For example, basic reading and writing skills are general skills that will not provide competitive advantage to any one firm. However, a firm that hired many employees who could not read and write would be at considerable disadvantage in the marketplace. Second, most organizations have defined the "New Deal" between the firm and it's employees. This new psychological contract (Rousseau & Greller, 1995) is characterized by employers assuring that they will not guarantee employment, but will guarantee employability to people (Kissler, 1995). This requires providing employees with the necessary training and development that ensures them marketability to other firms (i.e., general skills). Thus, firms that fail to invest in general skills will be unable to attract and retain competent employees.

In addition, while general skills are applicable across organizations, and thus most likely to result in only competitive parity, this does not preclude gaining competitive advantage through obtaining the highest level of general skills. For example, Wright et al. (1994) argued that firms that were able to obtain the highest level of average cognitive ability would have a competitive (and possibly sustainable) advantage. Thus, we would not argue for ignoring the importance of general skills; they add value and general skills at the highest level are rare.

However, greater potential for sustainable competitive advantage stems from investments in firm specific skills. One avenue to sustained competitive advantage is to focus on developing a firm specific skill base within an organization because these skills cannot be easily duplicated by competitors. These skills provide competitive advantage because they provide value to the firm, but that are not easily marketable by the employees who possess them. One can accomplish this through investing in constant training and development of employees to perform work processes and procedures that are specific to the firm. In fact, central to the concept of organizational learning is the process of developing and disseminating tacit knowledge (i.e., firm specific knowledge) throughout the firm (Miller, 1996; Senge, 1990). Thus, the firm gathers the rents accruing from these firm specific skills while providing employees with the opportunity for growth and development.

This highlights the potential shortsightedness of outsourcing most or all of a firm's training and development activities. Outsourced activities such as these most effectively provide general, rather than firm specific skills. While some training firms may be able to develop tailormade programs for specific firms, these are not feasible when proprietary technologies and processes exist. In addition, the training firm which develops the tailor-made programs consequently acquires the skills and can theoretically (although not ethically and possibly not legally) exploit them with competing firms. For these reasons, while some training activities can and should be outsourced, outsourcing of all training activities is not likely to serve as the lever for gaining sustainable competitive advantage through people.

Sustainable Competitive Advantage comes from Teams more than from Individuals.

Much of the popular literature on top management seems to point to individual CEO's such as Lee Iaccoca at Chrysler, Jack Welch at GE, or Lawrence Bossidy at Allied Signal as sources of sustainable competitive advantage. Similarly, much of the academic work on matching human resources to organizational strategies has focused on top managers and ignored the lower level employees (c.f., Gerstein & Reisman, 1983; Gupta, 1984; Gupta & Govindarajan, 1984; Guthrie, Grimm, & Smith, 1991; Guthrie & Olian, 1991). The inherent assumption in this research is that the skills of the workforce are all common across firms, but that highly skilled individual managers or top management teams are more rare (Wright et al., 1994). This implies that the firm that has the right CEO or President might possess a source of sustained competitive advantage. However, while these individuals are quite valuable, if labor markets are at all efficient, they are not likely a source of sustained competitive advantage.

Individuals who possess valuable and rare skills are usually able to claim most of the rents that are attributable to those skills (Wright et al., 1994). An outstanding Chief Executive, because of the high visibility of his or her performance, will soon be approached by other organizations with higher compensation. In the bidding process for that individual's services, he or she can claim most of the rents, and therefore the rents will not accrue to whichever firm finally obtains that individual's services.

Numerous recent shifts of top managers from one firm to another (e.g., Gerstner to IBM) as well as the rapidly rising top executive pay exemplifies the futility of seeking sustainable competitive advantage from the skills of one individual. On the other hand, the exploitation of the synergistic value from a large number of individuals who work together is quite costly if not impossible for competitors to imitate. Thus, teams or larger groups, due to causal ambiguity and social complexity, provide greater potential to be a source of sustainable competitive advantage.

Alchian & Demsetz (1972) defined team production as "production in which (1) several types of [human] resources are used and (2) the product is not a sum of the separable outputs of each cooperating resource," (p. 779). Because output is more than the sum of the separable outputs of each cooperating resource, it is difficult, if not impossible to identify the specific source of the competitive advantage. In other words, the competitive advantage stemming from team production is characterized as being causally ambiguous, thus, making it difficult for competitors to imitate.

An additional benefit of team production is that individuals become linked in transaction specific relationships, resulting in transaction specific human capital. In other words, team members become involved in socially complex relationships that are not transferable across organizations, thus only benefiting the organization in which these relationships develop. This requires the development of a team orientation, as has been exemplified among the top managers at Continental Airlines. One part of their turnaround was the replacement of 36 of the company's top officers in a 12-month time frame. CEO Gordon Bethune states "Why do you think most of those VP's disappeared? Most of them could not be team players." This has resulted in a reorientation among the top managers at Continental to focus on team goals instead of being strictly focused on their own personal goals (Boisseau, 4/22/95).

This highlights the importance of the HR function in developing and nurturing the relationships among organizational members. Many traditional Organizational Development activities such as team building and conflict resolution are now found in the HR departments of the Fortune 500 companies (McMahan & Woodman, 1992). In addition, researchers are beginning to explore trust among organization members as one determinant of firm performance (Mishra & Mishra, 1994). Clearly trust and good relationships among organizational members are firm specific assets that provide value, are quite rare, and are extremely difficult for competitors to imitate.

Sustainable Competitive Advantage Stems from HR Systems more than from Single HR Practices.

Much of the writing on Strategic Human Resource Management has focused on human resource practices as a source of competitive advantage (e.g., Schuler & MacMillan, 1984). The assumption is that firms that engage in the best human resource practices, i.e., have the best selection system, or best training program, or best reward system, etc., will have a competitive advantage over firms that fail to use this practice. Both the recent work on Utility Analysis of HR programs (c.f., Boudreau, 1991; Cascio, 1987; Jones & Wright, 1993; Steffy & Maurer, 1991) and empirical work on the relationship between HR practices and performance (e.g., Terpstra & Rozelle, 1993) has demonstrated that HR practices do provide value to the firm.

While each of these practices provide value, within the VRIO framework they are not likely to be a source of sustained competitive advantage alone. Given the recent emphasis on benchmarking to identify the most effective HR practices, any individual effective practice is easily imitated, and thus, can provide a temporary advantage only for a short time until competitors can copy it.

The fact that these individual practices will not likely lead to sustainable competitive advantage does not imply that these practices are unimportant and thus, HR executives can ignore identifying the best practice for each of the various HR activities. The failure to invest in state-of-the-art selection, training, and reward systems can result in a firm having a competitive disadvantage among human resources. In addition, a series of temporary competitive advantages gained through constant innovation is still quite valuable to the firm.

However, the challenge for HR is to develop systems of HR practices that create a synergistic effect rather than develop a set of independent best practices of HR (Becker & Gerhart, 1996; Gerhart, Trevor, & Graham, 1996; Lado & Wilson, 1994; Wright & Snell, 1992; Wright et al., 1994). This requires a changing mindset from the traditional subfunctional (selection, training, appraisal, compensation, etc.) view of HR to one where all of these independent subfunctions are viewed as interrelated components of a highly interdependent system. The interrelatedness of the system components make the advantage difficult, if not impossible for competitors to identify and copy. It also requires investing time and energy into developing systems and structures for integrating various HR practices such that they complement, rather than conflict with one another. While this sounds quite common-sensical, our conversations with a number of HR executives consistently indicate that very few HR departments have developed any such systems and structures. Thus, firms that have developed highly integrated systems seem to have obtained a source of sustainable competitive advantage. Recent research on bundles of HR practices supports this notion (e.g., Delaney & Huselid, 1996; Delery & Doty, 1996; Huselid, 1994; MacDuffie, 1995; Youndt, Snell, Dean, & Lepak, 1996).

These implications of the VRIO framework run against the grain of much of the current management thinking that emphasizes the importance of finding the right CEO, outsourcing HR functions, or seeking sustained competitive advantage through finding one best HR practice. It does not imply that these activities are not valuable, but only that they are incomplete, particularly in guiding the decision making of HR executives. The following section will examine implications of the VRIO framework for HR executives.

Implications for HR Executives

The VRIO framework presents a number of implications for HR executives. In general it highlights the fact that HR executives play an important role in managing the firm's human assets, those which possess the greatest potential for being a source of sustainable competitive advantage. However, more specifically, it provides guidance regarding the management of the HR function in organizations in ways that create competitive advantage. We outline four of these major implications below, with questions to help guide the HR executive in managing the function.

1. **Understand the value of people in the firm and their role in competitive advantage.**

Knowing the economic value of the firm's human resources is a necessary precondition before any HR executive can begin to manage the function strategically. Reichheld (1996) notes that people contribute to firms in terms of efficiency, customer selection, customer retention, customer referral, and employee referral. People play an important role in the success of any firm, but which people do so and how may vary across firms. This knowledge is a necessary starting point for any HR executive to act as a strategic partner.

For example, our research indicates that firms that rely heavily on innovation and product development (e.g., Merck) argue that their R&D scientist's ability to develop successful new products is the major thing that distinguishes them from competitors. Manufacturing firms such as Dell Computer, on the other hand, emphasize the production efficiency advantages they can gain through harnessing all of their peoples' skills and effort. Finally, service oriented firms such as Continental Airlines note that the planes, routes, gates and fares are virtually identical within the industry. Thus, their competitive advantage can only come through efficient, friendly service that makes fliers want to make their next flight on Continental.

Similarly, while all of the firm's people are important, some provide greater leverage for competitive advantage. Because of the need for innovation, Merck's R&D scientists provide greater leverage for success than do the hourly manufacturing employees. On the other hand, it is the hourly line employees (ticket agents, flight attendants, gate crews, baggage handlers) who directly impact the flying experience that have a relatively stonger impact on competitive advantage for Continental Airlines.

Thus, HR executives must first understand the role of the firm's people in competitive advantage before being able to make decisions about how to position the deliverables of the function. This leads to the following questions:

- On what basis is the firm seeking to distinguish itself from competitors? Production efficiency? Innovation? Customer Service?
- C Where in the value chain is the greatest leverage for achieving this differentiation?
- C Which employees or employee groups provide the greatest potential for us to differentiate our firm from our competitors?

Understand the economic consequences of the human resource practices in a firm.

Once an HR executive understands the specific ways in which the firm's people provide value, it is necessary to examine the value that HR provides or can provide. Recent research has uncovered a relationship between HR practices and the financial performance of firms (Delery & Doty, 1996; Huselid, 1995; 1996; McDuffie, 1995; Welbourne & Andrews, 1996; Youndt, Snell, Dean, and Lepak, 1996). While this research is promising, more research is needed on how, exactly, this impact is gained. We believe that there are two basic ways.

First, HR practices are important levers by which firms develop human capital and employee commitment. It is the HR practices that can directly impact the skills of the workforce that can provide value to the firm. These practices also can help to develop committed employees who are willing to allocate their discretionary behavior toward organizational ends (MacDuffie, 1995; Wright, et al. 1996). In other words, HR practices play an important role in developing the human assets that provide competitive advantage.

It is also important to understand that HR practices and the HR function incurs costs for organizations. Thus, HR can impact firm performance through its efficiency in developing the human assets that are a source of competitive advantage (Ulrich, 1997). The products and services provided by the HR function can be too many or too few, of high quality or of low quality, directly linked to business needs or unrelated to the business. For example HR practices developed because they are the latest fad without a careful analysis to their ability to meet.

strategic business needs are both excessive and inefficient. Similarly, the failure to develop practices that will help address business needs results in less than optimal organizational effectiveness. Or, finally, HR practices designed to meet business needs that are delivered at excessive cost or with low quality negatively impact the firm's financial performance. Thus, HR executives need to assess both the menu of HR practices and services offered as well as the quality and efficiency in their delivery.

For example, as part of Continental Airlines' turnaround, the HR function took a long look at what services they provided and how efficiently those services were provided. The result of this analysis was the elimination and consolidation of a number of training programs that were simply unrelated to the business while still keeping some of the remaining training programs internal to the firm, the outsourcing of benefits and some training/development activities, and the development of a variety of variable pay plans (the on-time bonus, management bonus plans, profit sharing, etc.) Even today the firm is exploring further outsourcing and strategic partnerships as ways to reduce the costs of the function. Finally, in an effort to remain close to their customers, the HR function recently surveyed the company's officers regarding the importance of the services provided by HR as well as HR's effectiveness at delivering those services. This effort will guide them on areas for further improvement.

Thus, HR executives seeking to explore the value created by their functions need to ask the following questions:

- Who are your internal customers and how well do you know their part of the business?
- Are there organizational policies and practices that make it difficult for your internal clients to be successful?
- What services do you provide? What services should you provide? What services should you not provide?
- How do those services reduce their costs/increase their revenues?
- Can those services be provided more efficiently by outside vendors?
- Can you provide those services more efficiently?
- Do managers in the HR function understand the economic consequences of their jobs?

3 **Understand how the human resources and human resource practices in your firm compare to those in competing firms.**

The previous two questions focus the HR executive's attention internally within the organization. However, in a competitive environment, one cannot ignore the actions of competitors, and this is also true of HR. Therefore, it is necessary to examine the HR functions of competitors to gain an understanding of what HR practices and relationships define the present competition. However, this information is only valuable so far as it is used for developing strategies for changing the competitive landscape to your firm's advantage.

Such benchmarking activity has become almost commonplace in industry as firms look both within and outside their industries seeking the "best practices." Benchmarking provides information that can be valuable or useless, depending upon how it is used. If the goal of the activity is simply to identify the HR practices of successful firms in order to imitate them, then the costs will likely outweigh the benefits. Benchmarking identifies the rules of competition in an industry, and can be maximally valuable in providing information on two issues.

First, it helps firms to identify what superior practices the competition is engaged in which might provide them a competitive advantage until other firms are able to imitate it. For example, 5 years ago Nieman Marcus, the upscale retailer, implemented a sophisticated applicant tracking system that significantly reduced their recruiting costs. Because the system was purchased from an outside vendor, it did not take long for competitors to imitate the advantage through implementing similar systems. However, had competitors not identified the system as an advantage, their financial performance might have suffered needlessly.

Second, benchmarking should be used to identify ways to leapfrog competitors. This is accomplished through developing innovative HR practices, and is especially successful if they are ones that competitors will find it costly or difficult to imitate. For example, one of Merck's manufacturing plants recently shifted to a variable pay system resembling a gainsharing type plan. This plan has been hugely successful even while others plants in the industry and geographic area have been disbanding such plans. Why did it work at Merck? Merck's manufacturing managers attribute the success to the fact that the company has traditionally had a culture that is characterized by high levels of trust between employees and management. Thus, the compensation system, while imitable in formulas, structures, and procedures, was not imitable in practice since its success was contingent on Merck's unique history and culture.

Thus, HR executives seeking to understand their functions in relationship to competitors as a means of identifying which practices should be copied to maintain competitive parity, which practices can be innovatively delivered to provide temporary advantage, or which practices can be linked to the unique situation (culture, history, other management systems, etc.) of the firm in order to gain sustainable competitive advantage. This leads to the following questions:

- How do the workforce skills of your competitors (particularly in key jobs) compare to those in your firm?
- How does the commitment level of your workforce compare to that of competitors?

- What are your competitors' HR functions doing in terms of practices and relationships with line managers? How can you beat them by doing things better or differently?
- What unique aspects of your firm (e.g., history, leadership, culture, etc.) might allow you to develop and/or maintain a more highly skilled and highly committed workforce?
- What HR practices need to be developed or maintained to exploit these unique aspects of your firm?
- Given your firm's history and culture, what unique HR practices might you be able to implement more efficiently and effectively than your competitors?

4. Understand the role of the human resources function in building organizational capability for the future.

A constant tension exists in the trade-offs between focusing decision making and resource allocation on the short and long term in most organizations. This conflict also exists within the HR function. Many HR functions are struggling so hard with meeting current needs that they have little time to explore long term organizational plans. However, this tendency must be broken if HR executives want to play the role of strategic partner.

For example, a high tech manufacturing firm we are familiar with has seen tremendous growth in both revenues and headcount over the past 4 years. This growth has resulted in the HR function struggling to keep up with the hiring and training needs of a firm growing by 40% per year. However, such growth also had made it difficult for the HR function to pay attention to developing the organizational infrastructure necessary to maintain the growth. Over the past 2 years the HR function in this firm has begun investing in developing the organizational capability through the creation a succession and developmental planning system for the management team and a human resource planning system for the rest of the organization. Without such an investment, the firm's growth prospects would be substantially limited.

Thus, in spite of the need to deliver the traditional HR services to meet the organization's current needs, HR functions must consider the future organization's needs through answering the following questions:

- What is the firm's core competence, or the core competence the firm is trying to develop in the next 5-10 years?
- What will be the competitive landscape 5-10 years from now in terms of your firm's product markets and labor markets?

- What kind of human resources will your firm need to compete successfully five years from now? Ten years from now?
- What types of HR practices are needed today to build the organization needed in the future?

CONCLUSION

One important implication of the VRIO framework is that the Human Resource function manages the set of resources (e.g., human capital skills, employee commitment, culture, teamwork, etc.) that are most likely to be sources of sustained competitive advantage into the next century. This should illustrate the increasing importance of HR in influencing organizational performance in today's competitive environment. However, in addition to highlighting the important role that the HR executive plays, it also provides guidance for the HR executive in how to effectively gain and maintain the role of strategic partner.

The VRIO framework helps the Human Resource Executive to evaluate all of the activities of the function against the criteria of value, rareness, imitability, and organization. Again, HR activities that are valuable but not rare, or valuable and rare but imitable are not to be ignored. These are the activities that the function must perform to maintain competitive parity or to provide temporary competitive advantages. For example, competitor firms can likely imitate a particular selection system that identifies cognitive abilities, technical skills, and/or interpersonal skills that provides value. However, to fail to identify these skills through selection can result in a severe competitive disadvantage.

The ultimate quest should be for the HR function to provide the firm with resources that provide value, are rare, cannot be easily imitated by other organizations. This entails developing employees who are skilled and motivated to deliver high quality products and services, and managing the culture of the organization to encourage teamwork and trust. This requires that HR functions focus more attention on developing coherent systems of HR practices that support these aims.

More importantly, however, is that the VRIO framework points to the need for an entirely new mindset regarding the role of HR executives in the organization. Many HR Executives complain that they have not been invited to the strategic planning "table." However, upon examination, it becomes clear that these executives either are unaware of or unable to clearly communicate to the strategic planners any economic reason that they should be at the table. It appears that there are far too many HR executives who view themselves as Human Resource people who happen to work in a business, than business people who happen to work in the Human Resource function. The VRIO framework enables business people in HR to transform the HR function into a contributor to firm performance rather than a drain on firm resources.

REFERENCES

1. Alchian, A. and Demsetz, H. (1972). Production, information costs, and economic organization, American Economic Review, 62, 777-95.
2. Barlow, J. (4/7/96). Clearing the air at Continental. Houston Chronicle, p. 1D.
3. Barney, J. (1991). Firm Resources and Sustained Competitive Advantage. Journal of Management, 17, 99-120.
4. Barney, J. (1995). Looking inside for competitive advantage. Academy of Management Executive, 9(4), 49-61.
5. Becker, B., & Gerhart, B. (1996). The impact of human resource practices on organizational performance: Progress and prospects. Academy of Management Journal, 39, 779-801.
6. Becker, G. (1964) Human Capital. New York: Columbia U. Press.
7. Boissueau, C. (10/22/95). Morale is higher as new managers and a return to profitability give workers a reason to have hope. Houston Chronicle, p. 1D.
8. Boudreau, J. (1991). Utility analysis for decisions in human resource management. In M. Dunnette and L. Hough (Eds.) Handbook of Industrial/Organizational Psychology, 2nd. Ed.. Vol. 2, 621-746.
9. Cascio, W. (1987). Costing Human Resources: The Financial Impact of Behavior in Organizations. Boston, MA: Kent.
10. Delaney, J., & Huselid, M. (1996). The impact of human resource practices on perceptions of organizational performance. Academy of Management Journal, 39, 949-969.
11. Delery, J., & Doty, H. (1996). Modes of theorizing in Strategic Human Resource Management: Tests of universalistic, contingency, and configurational performance predictions. Academy of Management Journal, 39, 802-835.
12. Flamholtz, E. (1985). Human Resource Accounting (2nd. Ed.) San Francisco: Jossey-Bass
13. Flamholtz, E., & Lacey, J. (1981). Personnel Management: Human Capital Theory and Human Resource Accounting. Los Angeles, CA: Institute of Industrial Relations, UCLA.

14. Gerhart, B., Trevor, C., & Graham, M. (forthcoming). New directions in employee compensation research. In G. Ferris (ed.) Research in Personnel and Human Resource Management.

15. Gerstein, M. & Reisman, H. (1983). Strategic selection: matching executives to business conditions. Sloan Management Review, 24, (2), 33-49.

16. Gupta, A. (1984). Contingency linkages between strategy and general manager characteristics: a conceptual examination. Academy of Management Review, 9, 399-412.

17. Gupta, A. & Govindarajan, V. (1984). Business unit strategy, managerial characteristics, and business unit effectiveness at strategy implementation. Academy of Management Journal, 27, 25-41.

18. Guthrie, J., Grimm, C., & Smith, K. (1991). Environmental change and the top management teams. Journal of Management, 17, 735-748.

19. Guthrie, J. & Olian, J. (1991). Does context affect staffing decisions? The case of general managers. Personnel Psychology, 44, 263-292.

20. Hashimoto, M. (1981). Firm-specific human capital as a shared investment. American Economic Review, 71, 475-482.

21. Hunter, J. & Hunter, R., (1984). Validity and utility of alternative predictors of job performance. Psychological Bulletin, 96, 72-98.

22. Huselid, M. (1995). The impact of human resource management practices on turnover, productivity, and corporate financial performance. Academy of Management Journal, 38, 635-672.

23. Jackson, S., & Schuler, R. (1995). Understanding human resource management in the context of organizations and their environments. Annual Review of Psychology, 46, 237-264.

24. Jones, G. & Wright, P. (1992). An economic approach to conceptualizing the utility of human resource management practices. In K. Rowland and G. Ferris (Eds.) Research in Personnel and Human Resource Management, 10, 271-299.

25. Kissler, G. (1994). The new employment contract. Human Resource Management Journal, 33, 335-352.

26. Lado, A., & Wilson, M., (1994). Human resource systems and sustained competitive advantage: A competency-based perspective. Academy of Management Review, 19, 699-727.

27. MacDuffie, J. (1995). Human resource bundles and manufacturing performance: Organizational logic and flexible production systems

in the world auto industry. Industrial and Labor Relations Review, 49, 197-221.

28. McMahan, G., & Woodman, R. (1992). The current practice of organization development within the firm: A survey of large industrial corporations. Group and Organization Studies, 17, 117-134.

29. Miller, D. (1996). A preliminary typology of organizational learning: Synthesizing the literature. Journal of Management, 22, (3), 485-505.

30. Mishra, A., & Mishra, K. (1994). The role of trust in effective downsizing strategies. Human Resource Management, 33, 261-280.

31. Penrose, E.T. (1958). The theory of the growth of the firm. New York: Wiley.

32. Quick, J. (1992). Crafting an organizational culture: Herb's hand at Southwest. Organizational Dynamics, 21, 45-56.

33. Reichheld, F. (1996). The Loyalty Effect: The Hidde3n Force Behind Growth, Profits, and Lasting Value, Boston, MA: Harvard Business School Press

34. Ricardo, D. (1817). Principles of Political Economy and Taxation. London: G. Bell.

35. Rousseau, D., & Greller, M. (1994). Guest Editors; Overview: Psychological contracts and human resource practices. Human Resource Management Journal, 33, 383-384.

36. Rumelt, R. (1991). How much does industry matter? Strategic Management Journal, 12, 167- 185.

37. Schmit, M., & Allscheid, S. (1995). Employee attitudes and customer satisfaction: Making theoretical and empirical connections. Personnel Psychology, 48, 521-536.

38. Schlesinger, L., & Zornitsky, J. (1991). Job satisfaction, service capability, and customer satisfaction: An examination of linkages and management implications. Human Resource Planning, 14, 141-150.

39. Schneider, B., & Bowen, D. (1985). Employee and customer perceptions of service in banks: Replication and extension. Journal of Applied Psychology, 70, 423-433. Schuler, R. (1992). Strategic Human Resource Management: Linking the people with the strategic needs of the business. Organizational Dynamics, Summer, 18-31.

40. Schuler, R. S. and Jackson, S.E. (1987). Linking competitive strategies with human resource management practices. Academy of Management Executive, 1, 207-219.

41. Schuler, R.S. and MacMillan, I. (1984). Gaining competitive advantage through human resource practices. Human Resource Management, 23, 241-256.

42. Schumpter, J. (1934) The Theory of Economic Development. Cambridge: Harvard University Press.

43. Senge, P. (1990). The Fifth Discipline. New York, NY: Doubleday.

44. Snell, S., Youndt, M., & Wright, P. (1996). Establishing a framework for research in strategic human resource management: Merging resource theory and organizational learning. In G. Ferris, (Ed.) Research in Personnel and Human Resource Management, 14, 61-90.

45. Steffy, B., & Maurer, S. (1988). Conceptualizing and measuring the economic effectiveness of human resource activities. Academy of Management Review, 13, 271-286.

46. Templin, N. (3/10/1994). Auto plants, hiring again, are demanding higher-skilled labor. Wall Street Journal, p. A1, A4

47. Templin, N. (12/15/1992). A decisive response to crisis brought Ford enhanced productivity. Wall Street Journal, A1, A13.

48. Terpstra, D. & Rozzell, E. (1993). The relationship of staffing practices to organizational level measures of performance. Personnel Psychology, 46, 27-48.

49. Tornow, W., & Wiley, J. (1991). Service quality and management practices: A look at employee attitudes, customer satisfaction, and bottom line consequences. Human Resource Planning, 14, 105-115.

50. Ulrich, D. (1997). Human Resource Champions, Boston, MA: Harvard Business School Press.

51. Ulrich, D., Halbrook, R., Meder, D., Stuchlik, M., and Thorp, S., (1991). Employee and customer attachment: Synergies for competitive advantage. Human Resource Planning, 14, 89- 104.

52. Waterman, (1994). Frontiers of Excellence, London: Nicholas Brealey Publishing.

53. Welbourne, T., & Andrews, A. (1996). Predicting the performance if initial public offerings: Should human resource management be in the equation? Academy of Management Journal, 39, 891-919.

54. Wernerfelt, B. (1984). A resource based view of the firm. Strategic Management Journal, 5, 171- 180.

55. Wright, P. M., & McMahan, G. C. (1992). Alternative theoretical perspectives for strategic human resource management. Journal of Management, 18, 295-320.

56. Wright, P. M. & Snell, S. A. (1991). Toward an integrative view of Strategic Human Resource Management, Human Resource Management Review, 1, 203-225.

57. Wright, P., McMahan, G., & McWilliams, A. (1994). Human resources as a source of sustained competitive advantage. International Journal of Human Resource Management, 5, 299- 324.

58. Wright, P., McCormick, B., Sherman, S., & McMahan, G., (1996). The role of human resource practices in petro-chemical refinery perform

59. Youndt, M., Snell, S., Dean, J., and Lepak, D. (1996). Human resource management, manufacturing strategy, and firm performance. Academy of Management Journal, 39, 836-866.

Chapter 6

Key Issues in Strategic Human Resources

Kate Walsh, Michael C. Sturman, John Longstreet

Cornell University, kmw33@cornell.edu

Cornell University, mcs5@cornell.edu

Quaker Steak & Lube Restaurants

ABSTRACT

This chapter, on strategic human resource management (SHRM), is really not for the hospitality human resources (HR) director. We already believe—whether through education and training, through insight gained from experience, or through self-interest— that HR directors accept as true the importance of an organization's HR function and its HR assets. The strategic nature of HR decision making extends far beyond the doors of the HR department. Just as money changes hands and is important to more than those in a finance department, nearly all decisions in a company's purview involve HR in some way, shape, or form. The purpose of this chapter is thus to help those already making HR decisions do so in a way that takes advantage of what we have learned from research in SHRM.

Keywords: hospitality management, hospitality industry, strategic human resource management, decision making.

This chapter, on strategic human resource management (SHRM), is really not for the hospitality human resources (HR) director. We already believe—whether through education and training, through insight gained from experience, or through self-interest— that HR directors accept as true the importance of an organization's HR function and its HR assets. The strategic nature of HR decision making extends far beyond the doors of the HR department. Just as money changes hands and is important to more than those in a finance department, nearly all decisions in a company's purview involve HR in some way, shape, or form. The purpose of this chapter is thus to help those already making HR decisions do so in a way that takes advantage of what we have learned from research in SHRM.

There are a number of critical challenges inherent in the nature of HR decision making that make its strategic implementation so difficult. Human assets have characteristics that differ from most other resources in the company, such as its physical product and financial capital. Most notably—and similar to other investments, such as the value associated with a hotel's name recognition or brand— they are intangible and therefore, simply unclear. Human performance is difficult to predict; it is very challenging to make a strong case that a particular investment in employees (e.g., a new pay plan, a new training program) clearly leads to specific improvements in employees' performance, which then has a direct effect on the financial results of a company. We look through a very fuzzy lens when we try to examine the financial value one person brings to an organization.

At the same time, costs associated with people are comparatively crystal clear. While the link from employee pay to employee attitudes to employee behaviors to customer perceptions to customer repurchase behaviors to organizational results is murky at best, the cost of that employee's pay is known with absolutely certainty, as is the direct cost of raising pay and providing training. Our traditional accounting systems add to this contradiction. Employees are not owned by their organizations and thus cannot be found on the asset side of the balance sheet. Instead, they only appear as expenses on the income statement. Because they represent a cost, it is tempting to treat employees as expenses to be minimized. The irony is that the work of people is one of the few organizational assets that can actually appreciate in value. The organizational leader making investment decisions about employees is looking through a very strange pair of bifocals indeed: Looking down, one is confronted by the blur associated with the value of intangible assets; looking over the top, one sees the clear image of their cost. Decision makers are confronted with the temptation of managing that which is easy to understand, especially when the results are immediately apparent. One of the goals of this chapter is to draw attention to the more long-term benefits that can be realized from investments in people.

Nowhere are these ideas more salient than in the hospitality industry. HR remains one of the key challenges hospitality decision makers face. Industry leaders in both the hotel and restaurant industries report human capital challenges—recruiting, retaining, motivating, training, and developing the workforce—as the problems that "keep them awake at night" (Enz, 2001,2004). Hospitality leaders are concerned because in service firms the success of products depends on their delivery by employees. Thus, service organizations rely on their employees to create memorable experiences that develop a loyal customer base and ultimately carry out the firm's strategic initiatives (Liao & Chuang, 2004; Skaggs & Youndt, 2004). Yet, most hospitality organizations operate with extremely lean margins. Many of these companies face triple-digit turnover and find it challenging to attract, retain, and develop a talented labor pool capable of creating relationships with customers that result in repeat business. At the same time, most organizations also are required to offer immediate, significant returns to their investors. Thus, when salaries and wages represent the number one expense item on the profit and loss (P and L) statement, decision makers find it difficult to justify increased investments in HR, as any additional pay or training budget cuts into immediate bottom-line profits.

The challenge decision makers face is thus twofold: How can organizations recognize the value associated with their human capital, and how can they make better decisions to manage the associated investments? While employee behaviors maybe difficult to predict, observe, and measure, investments in employees still need to be strategically managed. Drawing from current knowledge of SHRM, we help those making strategic decisions improve their insight to make choices about HR investments. We review the latest research in SHRM and examine how this research maps onto practical issues of concern in the hospitality industry.

The organization of this chapter is as follows: First, we discuss the challenges inherent in trying to understand and trace investments in employees and HR initiatives to the bottom line, and we consider these issues and challenges in practice. We review the challenge of understanding the impact of employee performance on organizational outcomes and discuss ways the use of HR metrics and analytics can aid in decision making. Following this discussion, the table is turned to the executive voice:

John Longstreet, former senior vice president of people strategy at ClubCorp and current president and CEO of Quaker Steak & Lube Restaurants, weighs in on the dialogue of HR strategy and offers some provocative insights and practices that he put in place at ClubCorp. In the last section, we reconcile different viewpoints between HR practice and research and discuss the implications for both constituencies, as well as for the field of SHRM.

CURRENT THEORY

SHRM is rooted in fundamental concepts in strategy, which essentially is about how to achieve firm profitability. Core research in strategy is concerned with identifying both the external and internal factors that enable firms to shape a competitive advantage and achieve superior performance (Harrison & Enz, 2005). To achieve superior performance, firms need to create cost efficiencies that enable them to operate on lean margins or alternatively differentiate themselves so as to charge price premiums. When examining competitive strength within the firm, the resource-based view (RBV) is helpful for identifying sources of advantage (Barney & Wright, 1998). This approach to strategy considers the role that internal resources, such as physical assets, organizational systems, and human capital play in helping a firm create value and become competitively distinct and profitable (Barney, 1991; Peteraf, 1993).

SHRM is concerned with two forms of resources. The first resource is an organization's human capital—the knowledge, skills, and abilities of its employees. The challenge for HR strategic decision makers is to transform human capital into a firm capability that is valuable, rare, and difficult to imitate (Amit & Shoemaker, 1993; Barney, 1991). The second resource is an organization's systems—specifically HR policies and practices—that serve to support the development of human capital. The challenge with this resource is to select the appropriate cost-efficient bundle that will help employees perform best.

Using these two forms of resources, SHRM is shaped by four primary challenges, or overarching goals. The first challenge is to tap into the intellectual and skill-based prowess of a company's employees and use this knowledge and skill set to the organization's best advantage. In other words, research in SHRM considers how to take employees as a source of human capital and turn this resource into a firm capability that creates competitive advantage. The second challenge is to turn an organization's HR infrastructure—or its bundle of HR initiatives and programs—into a source of competitive advantage—usually through setting procedures that enable a firm's human capital to appreciate in value. The third and overlapping challenge is to determine how a firm's HR initiatives can work in tandem to support the overall firm strategy. Central to this challenge is to understand the "black box" of return on investment (ROI) in HR initiatives, based upon the overall short-term and long-term financially based trade-offs that need to be considered. The final challenge is to apply HR tools, such as metrics and analytics, to aid in strategic decision making. We discuss each of these challenges in turn, including their implications for the hospitality industry.

Employees as a Source of Competitive Advantage

While a great deal of research in professional service firms, such as law and medical practices, focuses on the value employees bring to their organizations as human assets (i.e., Hitt, Bierman, Shimizu, & Kochhar, 2001; Kannan & Akhilesh, 2002; Sherer, 1995; Zucker, Darby, & Brewer, 1998), very little research has considered the value of employees performing low-skilled service work, such as those hired for hotels, restaurants, and other types of hospitality service organizations. With annual turnover, even of management staff, often climbing over 100%, the common line of thought and practice has been to curtail investments in employees, as they will likely soon leave the organization. This is despite abundant anecdotal accounts depicting the value associated with great employee service.

What we do know from research findings is that employees as a form of human capital have knowledge, skills, and abilities that can be applied to their work to generate "rents" or value for the organization (Becker, 1962; Bontis, 1998; Bontis, Crossan, & Hullard, 2002; Coff, 1997). When they are highly competent and performing their jobs well, employees are actually able to work in a tacit manner, meaning they are able to attend to a task at a skill level that is almost instinctive and does not require a great deal of planning or conscious thought. When a group of employees collectively act this way, they create an organizational system or routine that is so efficient it becomes a source of advantage or firm capability that other companies cannot replicate (Argote & Ingram, 2000; Coff, 1997; Hall, 1992; Lado & Wilson, 1994; Nelson & Winter, 1982; Nonaka, 1994; Teece, Pisano, & Shuen, 1997). One only has to think of a front desk clerk so competent working the computer system that he or she can seamlessly and simultaneously check-in—and graciously interact with—a hotel guest to picture what this type of performance can mean to a business. At the aggregate level, this application of human capital has been termed "knowledge value added" and refers to core organizational knowledge embedded within its routines and processes. It represents a collective and integrated use of employees' knowledge, skills, and abilities (Kannan & Akhilesh, 2002).

In addition to creating efficient routines, employees—especially those in service-based organizations—can create firm advantages by acting as boundary spanners that form crucial relationships with key stakeholders, such as customers and suppliers (Coff, 1997; Skaggs & Youndt, 2004). These sorts of links act as valuable sources of information that firms can use to shape their product and innovate. For example, employees can connect with restaurant customers in such a way that they can collect and pass along suggestions to the chef on how to create variation on the menu. The boundary spanner role is especially important to service organizations, because these sorts of customer interactions influence not only ways employees collect information but, simultaneously, ways customers help produce the service (Normann, 1984).

HR decision makers have long tried to argue that employees bring value to the organization and actually represent an investment that provides a substantial return. The questions are: What kind of return do they bring? What is the potential risk if employees leave the firm and bring their capital over to the competition (Coff, 1997)? Recent research has suggested intangible assets add approximately 50% to a firm's market value (Ulrich & Smallwood, 2005). As a beginning point in understanding the value of human capital as a type of intangible asset, researchers have found a positive relationship between the aggregated bundle of employees' knowledge, skills, and abilities and the organization's performance as reflected in profitability; however, as mentioned, most of this research has been conducted using professional service firms and more broadly, research-oriented industries, as a context (Bouillon et al., 1995-1996; Carmeli & Tishler, 2004; Hitt et al., 2001; Sherer, 1995; Zucker et al., 1998). With the exception of one exploratory study that offers preliminary ways to gauge human capital and customer capital in the hotel industry (and found an association between human capital and performance) (Engstrom, Westnes, & Westnes, 2003), this type of research is negligible in service firms that offer a basic product for sale. In reality, studies measuring the link between human capital and firm performance represent only a beginning point. The next question to logically consider is: What is the relationship between HR initiatives and firm value? In other words, do HR investments make a difference?

Human Resource Infrastructure as a Source of Competitive Advantage

What types of investments in employees yield the highest return? Specifically, what sorts of HR initiatives should organizations use to obtain the most beneficial use of employees' knowledge, skills, and abilities and as a result, create a competitive advantage? Following Huselid's (1995) work, a great deal of research emerged examining the impact of HR policy and practices on outcomes, such as organizational turnover, productivity, profitability, and market value (Becker & Gerhart, 1996; Delaney & Huselid, 1996; Huselid, Jackson, & Schuler, 1997; Wright, Smart, & McMahan, 1995; Youndt, Snell, Dean, & Lepak, 1996). The findings of one study suggest that one standard deviation change in an organization's HR program (termed "high-performing work system") can improve firm market value by $15,000 to $60,000 per employee (Huselid & Becker, 1995). At a minimum, companies with greater investments in their HR programs enjoy higher overall profitability (Becker & Huselid, 1998).

Similar to investments in human capital, researchers argue that investments in the HR infrastructure, such as hiring, training, and performance management systems, create "invisible assets" that enhance firm capabilities (Becker & Gerhart, 1996). These sorts of investments (1) teach and encourage employees to improve their knowledge, skills, and abilities; (2) motivate them to exert greater effort on behalf of their organizations; and (3) improve their overall productivity (Delaney &

Huselid, 1996; Youndt et al., 1996). They can include job rotation schedules, "empowerment programs, the use of participative decision making, and various forms of compensation strategies. In fact, the initiatives that appear to have the greatest impact on human capital are the presence of intensive and extensive training, competitive pay, and promotion-from-within policies. For organizations that rely on teams, the following HR programs foster the development of social capital and crucial knowledgesharing among members: socialization activities, mentoring programs, team-based reward structures, and incentives (Youndt & Snell, 2004).

Using resource-based concepts, Wright, Dunford, and Snell (2001) argued that the firm's core competencies are created when HR programs improve a company's human capital through influencing the creation, transfer, and integration of knowledge. In this framework, employees represent stocks of resources. HR programs can be used to increase not only human capital within an organization but also its level of social and organization capital (referring to the value of relationships among people and an organization's routines and processes) (Wright et al., 2001). Each of these "stocks" represents the creation, transfer, and integration of knowledge, respectively. Thus, HR programs are the processes by which a firm creates a combined flow of knowledge that represents a valuable, rare, and inimitable competency. The real advantage comes from treating HR programs as dynamic approaches shaped by a firm's overall strategy. When HR programs are shifting and interdependent, it is difficult for competitors to copy them (Becker & Huselid, 1998; Wright et al., 2001).

When they are effective, HR initiatives work in tandem to create a competitive advantage, such that the effect is both direct and multiplicative (Delaney & Huselid, 1996). Firms gain a competitive advantage not only because they undertake HR programs but also because of the way they selectively choose the appropriate mix of programs (Wright et al., 1995). While we know that a well-designed program of HR initiatives is strongly related to an organization's value, we still understand less about the ways in which they work together to do so. In other words, researchers are now trying to understand the black box of HR investments (see Exhibit 27.1). Specifically, how does the investment mix of HR initiatives help improve a firm's depth of human capital such that the company creates a unique competitive advantage, reflected in enhanced firm performance? What is the process by which this happens? (Becker, Huselid, Pickus, & Spratt, 1997).

Identifying and Opening the Black Boxes: From the Strategic Mix to Customer Impact

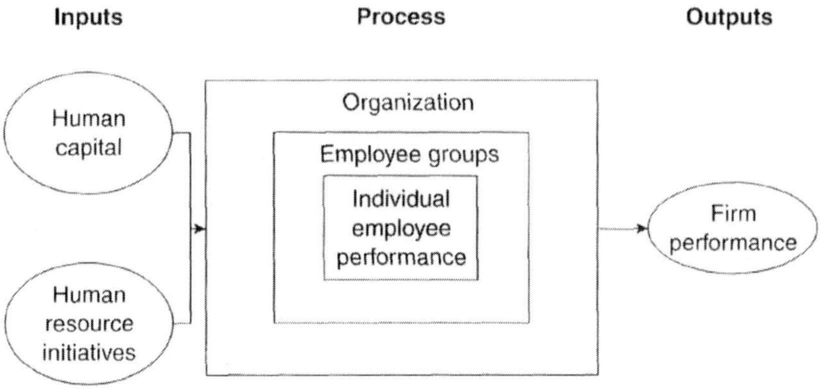

Figure 1 . The Black Box of Human Resources

As one can surmise by reading this chapter so far, while SHRM research has identified appropriate strategic questions and problems, the field is still a long way from providing definitive answers for decision makers to use. In part, this is because of the various approaches researchers take to examine strategic HR issues. Different areas of research in SHRM have treated different phenomena as "black boxes"—that is, researchers have identified inputs and outputs to these boxes and made conclusions about the effects of the inputs based on the associated outputs. Unfortunately, researchers still cannot explain the process by which inputs (i.e., human capital and HR programs) influence outputs (i.e., firm performance); hence, the internal workings of the black box remain a mystery.

The early work by Huselid and Becker (1995) on high-performance work systems treated companies as a whole, as a black box. These researchers examined the HR-related inputs to the box, identified as high performance work systems, considered the associated outputs, and drew conclusions about the relationship between the two (e.g., Delaney & Huselid, 1996; Huselid, 1995; Huselid et al., 1997).Their work leads to the premise that greater use of high-performance work systems is related to better business outcomes. Unfortunately, this work is not definitive in identifying and explaining what goes on inside the "black box" of the company.

In addition, a central problem with treating the entire company as a black box is that the concept implicitly assumes that all employees should be treated similarly—that is, one could conclude that high-performance work systems should be used with all employees to reap the potential benefits. As Boudreau and Ramstad (1997) noted, however, it is naive, and definitely not strategic, to think about managing all HR in the same way. Companies do not invest in all properties in a similar manner; not all products are marketed the same way and not all customers are treated identically.

Similarly, HR investments should be selectively managed using a decision science process that Boudreau and Ramstad (2005) termed "talentship." Which employees or groups of employees provide the highest return to the organization? What types of HR investments should be made in these employees?

Questions such as these prompted research viewing different employee groups as the black box to be opened and understood. Researchers who study the HR architecture examine the appropriate strategy to employ with different employee groups (Lepak & Snell, 1999; Youndt & Snell, 2004). These researchers argued that employees vary with regard to their uniqueness and strategic value, making some employee groups more valuable to the organization than others. Applying terminology from the RBV, uniqueness refers to the degree to which the employees work is rare, specialized, and/or firm specific (Lepak & Snell, 2002). Strategic value refers to the ability of the employee to "improve the efficiency and effectiveness of the firm, exploit market opportunities and/or neutralize potential threats" (Lepak & Snell, 2002, p. 519). Employee groups that fall into both categories require different HR investments than, for example, employees that fall into neither.

To create HR capabilities that are valuable, rare, and difficult to imitate, decision makers need to consider how to package HR initiatives to realize employees' potential (Barney & Wright, 1998). The real advantage comes not from initiating HR programs but rather, from actually bundling them together for different groups of employees in coherent, strategic ways (Lepak & Snell, 2002; Wright & Snell, 1991). Thus, the inputs to the black box are customized through different HR initiatives such that the organization creates a unique competency and competitive advantage (Jackson, Schuler, & Rivero, 1989).

Research on the HR architecture would suggest that employees, such as hotel employees, have strategic value to the firm but they are not necessarily unique; their skills are widely transferable. Thus, this employee group should be viewed with a short-term orientation. They should be paid market wage, evaluated using a results-oriented measure and managed with the possibility they may leave. More developmentally, long-term HR initiatives, including using above-market compensation plans as well as goal-oriented appraisal systems, should be applied to employee groups whose skills are more unique and firm-specific, such as knowledge-based workers, including perhaps the management staff of a hospitality organization (Lepak & Snell, 2002). However, the more a hospitality organization competes by trying to differentiate its level of service, and the more that employees are expected to utilize company-specific knowledge (e.g., for handling customer complaints, for providing exceptional service), then the more that these companies should invest in these employees, enhance their human capital, and strive for retention.

Treating groups of employees as a "black box" has problems similar to that of treating all employees the same. Even employees within a strategic group respond to stimuli in different ways, have various preferences, and perform in their unique ways. In response, some researchers have treated an individual employee's performance as the black box (Boudreau, Sturman, & Judge, 1994), thus prompting a stream of research termed "utility analysis" (see, for example, Boudreau, 1991, for a review). Utility analysis estimates the value of a HR intervention by first considering the effects of an HR program on employee performance and then by determining the value of the program through an estimate of the value associated with improved employee performance. A more thorough discussion of this method is presented later in the chapter, but the primary concept is that it is possible to quantify the additional benefits employees bring to their organizations through HR interventions. The way that improved performance is converted to a dollar value, though, has long been viewed as one of (if not the most) controversial aspect of this research (Boudreau et al., 1994). Thus, while having a better understanding of how HR programs affect individual employees may be a more useful and accurate way of examining the inputs associated with HR programs, the difficulty of quantifying the outputs associated with the better performance has made the utility analysis tool less than successful in practice.

Finally, one can go beyond the utility analysis perspective and still open the black box of employee performance. Instead of considering employee performance as an outcome that has some associated value, one could specifically look at how HR programs operate, through the employee, to affect outcomes associated with business success. Specifically, how do HR programs influence customer outcomes? Rather than match HR initiatives with the appropriate employee group or focus on stocks of capital and flows of knowledge some researchers argue that all HR initiatives should be focused on meeting not employees' needs but rather, customers' needs.

This concept maps well onto the service-oriented nature of the hospitality industry. Because service quality is so crucial to a hotel, restaurant, or travel firm's customer base and, relatedly, its customer management programs, HR initiatives can add value when they help improve ways employees connect with customers and meet their needs (Kundo & Vora, 2004; McGovern & Panaro, 2004). HR programs, such as staffing and training, can and should be viewed from the customer's perspective (Ulrich & Brockbank, 2005). In other words, decision makers can ask themselves: What would our customers think of the employees we hire? What types of training would be important to our customer base? How do we measure improvement in our customer satisfaction levels or in repeat customer business from these programs? Questions such as these help focus HR decisions on the key product for sale and keep HR decisions aligned with the organization's overall goals (Ulrich & Brockbank, 2005).

Unfortunately, while there is certainly conceptual appeal associated with relating HR programs to customer outcomes, it, too, is a path fraught with obstacles. On one hand, decision makers could again treat the entire company as a black box and examine the association between HR programs and customer outcomes; yet, once again they are confronted with the lack of knowledge as to what really is going on because of the HR programs. On the other hand, if decision makers try to research the links from HR programs to employee behaviors to customer outcomes, they are faced with a daunting research task. The field of HR needs to develop better tools and methods if it is going to make investigations such as these possible.

With all of these approaches to studying the strategic value of HR, researchers are trying to provide information to help companies understand how HR practices affect companywide outcomes. But the field has far to go, and the insights gained from any one perspective must be viewed cautiously, as each method has its own set of assumptions and flaws. We therefore must consider what companies can do on their own behalf to advance their understanding of strategic HR value and how they can make better decisions in this regard.

Human Resources at the Strategic Table

Although many argue that HR decisions have important strategic implications, because decision makers are having difficulty quantifying these implications, as they can more easily do with other types of tangible investment choices, they are not making much progress toward using HR data in meaningful ways (Lawler & Mohrman, 2003). Yet it pays to be strategic, and the vagaries of HR outcomes are not impenetrable obstacles. As an example, consider the field of marketing. Similar to HR, marketing deals with unclear decision paths, psychological phenomena, and hard-to-predict outcomes. Yet the field of marketing has developed the tools and analytics to model customer behavior and help decision makers make better strategic choices. The difficulties inherent in HR decision making present challenges, but the field can follow the model of marketing and develop the methods and analytics (Boudreau & Ramstad, 2005). The benefits of this course of action are becoming evident. The SHRM literature suggests that those who can manage their HR more strategically will be rewarded with greater financial returns (Becker & Huselid, 1998).

Yet as mentioned, while organizations, especially service-based firms, have been quick to cite employees as the most important asset, when expenses need to be minimized, many HR initiatives, such as training programs, are the first to be reduced or eliminated (Barney & Wright, 1998). One reason for this reaction is that many HR professionals lack the type of analytic and data-based decision-making capabilities needed to understand and communicate the return associated with their investments (Lawler, Levenson, & Boudreau, 2004). Thus, to manage HR investments strategically—similar to ways companies manage their financial assets and their products—it is necessary to gather information and use appropriate analytics to guide decision making. This requires the development of HR metrics and the use of HR decision-making models. We review some basic tools at hand.

HR METRICS

Boudreau and Ramstad (2003) identified three categories of metrics that can aid in the understanding and evaluation of HR investments: efficiency, effectiveness, and impact (see Exhibit 27.2 for a summary). Metrics of efficiency ascertain how well the HR function performs its basic administrative tasks and include items such as absence rate, vacancy rate, time-to-fill positions, cost per hire, training cost per employee, etc. These metrics are the easiest to collect and provide useful information on the effectiveness of HR administration. Regrettably, they do not provide insight into how HR practices help improve organizational performance.

The second kind of metric captures effectiveness. These metrics help reveal if HR programs have the intended effect on the people whom they are supposed to influence. For example, training programs should be evaluated on more than just participation or cost (efficiency measures); they should be evaluated on the sort of capabilities (i.e., knowledge, skills, abilities) gained by participants. Staffing programs should be evaluated by the effectiveness of new hires provided by the program. Whereas an efficiency measure of staffing may consider time to hire and turnover rate, effectiveness measures should capture the quality of stayers versus leavers, if dysfunctional turnover is decreasing, and if the hiring system is providing employees with the competencies needed to deliver the company's service or product.

The third type of metric helps determine if HR systems are developing and optimizing the capabilities of the company. Lawler et al. (2004) described this sort of metric as allowing one to demonstrate "a link between what HR does and tangible effects on the organization's ability to gain and sustain competitive advantage" (p. 29). Ultimately, the "holy grail" for HR management is to demonstrate the effectiveness of HR programs on the bottom line. The creation of this sort of metric, though, is not a simple process. Relevant numbers cannot be easily looked up in an information system or captured through an employee or customer survey. Rather, one must employ statistical techniques and/or experimental approaches. It is only through this sort of research-based approach, a process we label here human resource analytics, can a decision maker attempt to understand the effects caused by the implementation of HR programs.

	Type of Human Resources Metric		
Human Resources Program	Efficiency	Effectiveness	Impact
Staffing	• Time to hire • Cost per hire • Turnover rate	• Performance level of hires • Turnover rate change	Increase in customer repurchase intentions due to new staffing program
Training	• Training cost per employee • Training program participation	• Skill development achieved by training program	Increase in customer satisfaction due to new training program
Compensation	• Payroll costs • Compensation/ revenues	• Ability to hire top choices • Retention of top performers	Decrease in complaints due to new performance recognition program

Exhibit 2. Examples of the Different Types of Human Resources Metrics

Human Resource Analytics

To understand HR analytics, the comparison to the field of marketing again provides a useful illustration. Marketing research is an entire field devoted to developing methods for providing information that can guide marketing decisions. The material is commonly taught as part of a basic education in marketing. Those obtaining degrees in marketing learn about conducting experiments, surveying customers, and using statistical analyses to estimate the affects associated with given programs. Unfortunately, there is little comparable formal "science" in the HR field.

In part, the necessary tools are already present. The fundamental principles behind marketing research can be applied to understanding HR programs. Experiments, surveys, and analyses can be used in conjunction with relevant HR metrics to guide decision making. Just as, for example, an advertising campaign can be evaluated in terms of how well it attracts new customers, the same techniques can be applied to evaluate the effectiveness of a recruiting campaign for job openings; the methods used to evaluate the promotional campaign can be used to assess a new pay-for-performance system. The experimental and analytical techniques employed in marketing research can be used to estimate the affects associated with any number of HR programs, whose effects are captured by the effectiveness and impact metrics discussed earlier. These concepts are beginning to be applied in the training evaluation literature that examines the ways employees' improvements with their job skills (following a training intervention) are tied to improved organizational performance and profitability.

In addition to borrowing techniques from marketing research, there have also been advances in HR analytics specific to the HR field. Specifically, cost-benefit analysis tools have emerged to help guide HR decision making. As mentioned, utility analysis is a tool designed to quantify the benefits associated with HR interventions. Based on the assumption that individual performance has some quantifiable dollar value, utility analysis presents a set of techniques to estimate the benefits and costs associated with specific HR programs (for an introduction to the utility analysis tool, see Sturman, 2003). In brief, the utility analysis tool calculates the impact of an HR program by multiplying the number of employees affected by a program by the impact of the program on individual employee quality improvement. An estimated cost associated with the program is subtracted from this number (Boudreau & Berger, 1985). While the basic technique is fairly straightforward, research developing the tool has made refinements of the technique more accurate, as well as more complex (Sturman, 2000). Specifically, it is often difficult to estimate specific improvements an HR intervention has on an employee's performance

The potential for sophisticated HR analytics has existed for several decades. The utility analysis tool has existed since the 1950s but is still generally unused by managers. Some have argued that this is a flaw associated with the technique itself (e.g., Latham & Whyte, 1994; Whyte & Latham, 1997). Others have argued, though, that the sophistication of the method combined with little formal training on its use limited its widespread adoption (Sturman, 2000). Similarly, while the field of marketing research goes back to the 1920s (Bartels, 1965), there has not been any sort of widespread application of HR analytics as a component of HR education. Nonetheless, it is our hope that such approaches to HR decision making can become a common element in business education.

In sum, we argue that individuals making strategic HR decisions should be trained in the use of HR analytics to guide that decision making. To obtain the most from a firm's human capital and HR, decision makers need to be able to make good HR choices. Whoever is making decisions regarding the investment in, and deployment of, an organization's human capital needs to have the necessary information and (generally, computerized) tools to assist in analysis.

Issues in Practice

SHRM research provides many general suggestions on how to be more strategic or at least on how to approach HR decision making from a more strategic perspective. Lacking, though, are specific recommendations that have immediate applicability for practice. This is not surprising, nor should it be expected, as the specifics of every business will necessitate different approaches. The purpose of this chapter is not to provide the one path to strategic HR success; rather, our goal is to help readers learn from the advances in the field so that they can strike out on their own path. We offer a basic trail guide—ideas to consider when linking employees with firm strategy.

From an applied perspective, the first key piece of information needed before HR investment decisions can be made is to understand the value associated with employee performance. In an industry focused on the delivery of customer service, there are very few examples of where employee performance does not matter. Instead, the question becomes: How does employee performance make a difference? The more it matters, the more decision makers should examine their HR investments. Investments may be needed to choose the right employees (e.g., invest in a selection system and attract with appropriate salary and benefits), develop the employees that make the biggest difference (e.g., invest in training), and retain these employees (e.g., invest in their pay), as well as collect information on these employees (e.g., invest in performance appraisal systems). The greater the value of employee performance, the more that interventions designed to improve performance can produce value. Thus, the more value there is associated with employee performance, the more a company may want to invest in this human capital.

Decision makers also need to consider what resources are needed to make strategic HR choices. Companies must consider whether they should "make or buy" their HR as part of an overall competitive strategy. HR decisions, and even human assets, can be outsourced. For example, when selecting a housekeeper, a hotel can fully handle the hiring process from recruitment to evaluation to selection; alternatively, they can hire agencies to manage the selection process. Then again, they can simply outsource the entire function and thus hire another company to perform all cleaning tasks. This "make or buy" decision must be made very carefully. If one does not consider the potential value associated with employee performance, then the decision will be driven by a cost perspective. The collection of important HR metrics, combined with a scientific approach to HR analytics, can help determine the most profitable decision.

Drawing upon a hospitality analogy, so to speak, the strategic management of HR is like playing blackjack. Predicting any single behavior or result, like predicting any single hand dealt, is difficult at best. One could have a perfect photographic memory, a salient understanding of probability, and the ability to cognitively compute all the necessary numbers; but, if one sits down and plays a hand with no prior information, the chances of winning are nearly no better than the intuitive (drunk) player sitting at the next seat. But, as one collects information, plays hand after hand, and makes wise decisions as to when to "invest," the chances of winning increases. For SHRM, while developing the right system is difficult, the house is not nearly as stacked against the decision makers. If decision makers draw upon the existing research, including their understanding of human capital, and build a solid HR system, companies using a strategic approach to HR decision making can win—and win big.

COMMENTARY

By John Longstreet

Much of the debate regarding SHRM revolves around the ROI of HR initiatives. If we assume that, as a rule, HR initiatives are designed to, in some way, improve the "employee experience" at work, the justification for a considerable focus on HR is found in The Service Profit Chain by Sasser, Schlesinger, and Heskett (1997). The book focuses on the relationship between employee satisfaction, customer satisfaction, and profitability.

Ten years before the book was written, as a hotel general manager (GM), I embraced a concept that if I took care of my employees, they would take care of my guests, and my hotel would be successful. Certainly appropriate facilities, ongoing capital improvements, sales and marketing, cost control, and a myriad of other details are important. But it was crystal clear to me that if my employees were happy, as a rule so were my guests, and our gross operating profit (GOP) exceeded other operations without a similar "employee focus."

In 1998, with the assistance of Cornell Hotel School researchers, the service-profit chain was "proved up." Looking at 76 hotels, 6,500 employees, and a representative sample of the guests who stayed at those hotels, the researchers were able to establish a statistical correlation between employee satisfaction scores, guest satisfaction scores, and "GOP to budget" comparisons (Simons, 2002). A similar conclusion was reached in a 2005 analysis of 160 businesses and country clubs, their employees, and members at ClubCorp.

So, if we will agree that themes expressed in the service-profit chain are valid, the next question—logically—would be "What HR initiatives are most important in driving employee satisfaction?" I would argue that the best way to answer this question is to ask the employees. Combining our own data with that which has been collected from employees for years and compiled by organizations like the Corporate Leadership Council, clearly a huge component of employee satisfaction was, not surprisingly, "their supervisor." So, at ClubCorp we embarked on a program, designed in simplest of terms to improve our supervisors and managers. The program, called Leadership 101, was introduced in 3-day sessions held around the country to some 1,200 managers and supervisors. This is not to ignore other key indicators of satisfaction, like compensation, benefits, and recognition, but having a team of supervisors equipped to create a great-place-to-work environment seemed like the most logical first step.

This mission of "creating a great place to work" should be the overriding mantra of the HR function. That said, it is the operators who actually impact the work environment on a day-to-day basis, and good HR programs cannot make up for bad managers. Consequently HR's role is as the cheerleader for the cause and as a resource to assist operators in the achievement of this goal. Using a customer service model, the employee is the "customer" of the manager and supervisor, and the manager and supervisor is the "customer" of the HR department. Of course, another key role of HR is to help ensure we find and hire the right supervisors and managers in the first place.

The concept of HR as a service provider to operators leads to another key theme among HR circles and mentioned earlier in this chapter. How can HR professionals get the attention of operators or, put another way, how can they get a "seat at the table"? As an operator turned HR person, I have found that firsthand knowledge and empathy of "ops" has been absolutely critical to gain buy in from the operators on strategic HR initiatives. Operators are more likely to react well to terminology that they understand and conversely tend to bristle at "HR speak." As in reaching any audience, it is important that you speak their language.

Besides determining what HR initiatives are most important and how to gain acceptance, the chapter also deals with identifying employee groups that "provide the highest return to the organization" and tailoring HR investments accordingly. While acknowledging that training programs must differ based on the level and position of the employee, we reject that notion that HR programs should be different for one group of employee versus another. One of the most powerful leadership models, in my opinion, is "side-by-side" leadership, recognizing that every employee, manager, and executive works side by side in achieving a common company vision. In a hotel, if there were employees more important than others, it might be the room attendants or dish machine operators, since these are the hardest positions to fill and perhaps the hardest jobs to do. Side by side leadership is important in maintaining happy employees.

In summary, the service-profit chain provides the framework by which HR priorities and strategies should be developed. HR needs to recognize operators as their customers, understanding and empathizing with their challenges and needs. In that way, HR is more likely to gain a "seat at the table" and gain acceptance of important strategic initiatives.

Point-Counterpoint

We very much appreciate Mr. Longstreet's comments and ideas; readers should know Mr. Longstreet successfully led an extremely effective HR program at ClubCorp. He is now president and CEO of Quaker Steak & Lube Restaurants. Research has a great deal to learn from business practices that work, and his HR strategic initiatives at ClubCorp are a model to follow. We are grateful for the insights he offers to this chapter. With the goal of learning from dialogue, we'd like to engage in one here and offer a counterpoint to some of Mr. Longstreet's arguments. We specifically focus on the rub between research and practice and identify areas both where we disagree and agree with ideas being applied in industry practice.

From a perspective of those who are working with HR's impact and challenges in real time, we agree with Mr. Longstreet that there is a lot of appeal to the service-profit chain, especially the notion of asking managers, as well as HR, to focus on improving the employee experience. These ideas are akin to more recent ideas in the leadership literature of servant or super leadership (Manz & Sims, 1991). While we do not dispute that it is best to treat your employees as customers, and in doing so create a safe and desirable work environment, we do believe that the service-profit chain has the potential to be somewhat limited in its overall usefulness, mostly because of the employee satisfaction-performance link.

While the relationship between job satisfaction and profitability has been substantiated at ClubCorp, as a general notion, the link between job satisfaction and employee performance often proves tenuous. Research continues to suggest that a happy employee is not always high performing (Way, Sturman, & Raab, in press). For example, in an analysis of 312 studies, with a combined sample of over 54,000 employees, the estimated relationship between job satisfaction and job performance was around 0.30 (Judge, Thoresen, Bono, & Patton, 2001). Or in other words, less than 10% of the variance in job performance can be predicted by an employee's level of job satisfaction. (In addition, the causal link between satisfaction and performance is not entirely clear. For example, a good case can be made that if employees are performing well on the job and are recognized for that, they will be happier employees.) While the relationship between the two phenomena is still positive, and focusing on job satisfaction is arguably a very important endeavor, there may exist additional ways for HR to contribute to business success.

Yet, as we'd suggest to Mr. Longstreet, we may just be telling the same story using different words; the "side-by-side" leadership program Mr. Longstreet describes represents a valuable investment in the company's human capital. Both practice and research would agree that the goal of strengthening leadership within organizations is crucial to employees having the ability, motivation, and opportunity to perform well. In fact, Schneider and colleagues (2005) recently found a causal link among servicefocused leadership, employees' use of organization citizenship behaviors with customers, customer satisfaction, and sales. Establishing these types of links between leadership, employee performance, and critical measures of success such as sales and employee turnover, may make clear how HR can help drive organizational success.

We believe Mr. Longstreet to be dead-on when he asserts that HR needs to understand the nature of the business and—"must have firsthand knowledge and empathy of operations"— to have the credibility, and hence potential, to make meaningful changes to the workplace. Keith Hammonds recently made quite a stir in the HR field with a paper entitled "Why We Hate HR" (Hammonds, 2005). Part of the problem with HR management that Hammonds identifies is exactly this issue. HR managers need to know who is the company's core customer; they need to know who is the competition, and they need to know what the company does well. This requires knowledge beyond just what is managed within the walls of the HR department; it requires knowledge of the business. It also speaks to one of the four core competencies identified by Ulrich (1997) as crucial to HR's ability to impact its organization .

Training programs aside, Mr. Longstreet rejected our notion that HR programs should be different for one group of employees versus another. In the spirit of good debate, here, we simply have to offer a counterargument. While we agree that employees should uniformly be treated ethically, legally, and with respect and dignity, as expressed by Hammonds (2005), sameness is not a proxy for equity. It is not simply that different jobs contribute to the organizational mission in different ways (as Mr. Longstreet implies when he comments about the critical role of the frontline employee), but also that employees within that job perform at different levels. HR decision making needs to be all about differentiation. Which specific employees need specific skills? How can managers ensure that they retain their top talent? Should top performance be encouraged with pay increases that are among the highest in the organization? How can we encourage average or below-average employees to either increase their performance or perhaps leave the company? Again, our disagreement here is likely a matter of semantics

Clearly, we are in agreement with Mr. Longstreet that HR needs to contribute to business success, and perhaps we only differ slightly in the mechanisms through which this happens. Yet, this potential disagreement actually highlights one of our other major points in our chapter: HR needs to be able to assess its impact and understand how HR decisions (both at the policy level and at the individual level) make this happen. While the service-profit chain has proved valuable for ClubCorp in identifying a relationship between the "black box's" inputs and outputs, we still needed metrics and analytics to learn more about this issue and explain the process by which value-added occurs. The SHRM field can help here, as can greater researcher-practitioner collaboration. We hope this chapter encourages work in this area.

Reconciliation and Insights

As both research and practice have shown, the business of managing an organization's investment in its human capital is complex. Clearly, one of the central goals of HR decision makers is managing this investment portfolio through assessing the return of its bundle of HR initiatives. We emerge from our discussion of the process of SHRM with a salient, overarching theme.

HR decision makers will not be effective in managing their HR investments without delineating specific ways HR initiatives contribute to a firm's overall competencies, goals, and objectives. While a seemingly simple concept, as we hope we made clear, there exist a variety of ways to approach this goal—none of which are easy to prove. Thus, to even engage operational leaders in the discussion, HR decision makers need to apply the language and metrics that will make all organizational leaders take notice of ways their human capital appreciates in value and, to borrow concepts from the RBV, shape a firm's distinguishing competencies. This includes, as Mr. Longstreet so aptly pointed out, understanding the business' needs and practices, in addition to understanding the industry in which the business operates. With this information in hand, HR decision makers can engage operational leaders in shaping the organization's competitive advantage through reinforcing cultural values and developing employees' knowledge and skill-base. They can also help enact change and create organizational agility through aligning the hiring, training, compensation, and firing processes (Brockbank, 1999). With all of these processes, effective HR decision makers should use business core competency goals and metrics as the benchmark for evaluating HR initiatives (Ulrich, 1997; Wright, Snell, & Jacobson, 2004).

As Brockbank (1999) suggested, HR leaders need to go beyond the logical role of matching HR processes to a firm's strategy and be what he terms "strategically proactive." This means that HR leaders should be focused on tomorrow's organizational challenges. They should work on enhancing an organization's innovative capabilities through removing obstacles to creative processes and streamlining ways operational decisions are made. HR leaders also need to be crystal clear on what their organization knows (knowledge), what it does (skills), and how it thinks (culture). In doing so, they will be able to contribute more effectively to organizational change plans, including mergers and acquisitions (M & As). Finally, HR decision makers need to be in continuous conversion with sales and marketing personnel, staying ahead of customer trends and understanding ways employees can help attract and sustain a loyal customer-base (Brockbank, 1999).

This concept of linking employees with customers leads to our final insight. We believe the area of greatest potential for service organizations lies in understanding the crucial ways employees create value for the purchasers of the organization's product. More recent research in marketing examines the lifetime potential revenues customers bring to an organization, as well as the value of maintaining customer relationships (Duffy, 2000; Gupta &

Lehmann, 2003). It is widely understood that frontline service employees play crucial roles in the reproduction of intangible service experiences or products (Bowen, 1986). The wild cards in the equation are the customers, who often are not even cognizant of their own service expectations let alone their satisfaction with service experiences—or their own role in shaping those experiences. We understand that service encounters are complex and variable and as a result, often difficult to script.

What we need to understand are the financial ways employees contribute to the lifetime value of loyal and repeat customers. Broadly, we know that employees can contribute through communicating customer needs to preferred-customer databases, to learning various ways customers wish to reproduce service experiences, and to being such an efficient and recognizable part of the service experience that loyal customers are willing to pay price premiums for the ease and predictability of the service encounter (Reichheld, 1996). These types of behaviors turn what is termed "encounter-based" service experiences into "relationship-based" service experiences that add value to the firm. The customer switches from doing business with an organization to doing business with people who represent the organization (Gutek & Walsh, 2000). Research suggests that the customers who form relationships with their service organizations actually appreciate in potential value, quite similar to the investments in employees (Reichheld, 1996). What we do not know is how to measure the specific ways employees contribute to the long-term potential of an established customer base, as well as help attract a new one. Numerically linking these contributions to long-term revenue brings home the crucial ways employees contribute to any successful service business. Human capital—-and its associated HR investments—represent intangible assets that have remarkable potential to enhance both current and future firm value.

Implications for Strategic Human Resource Management in Hospitality

What do these insights suggest for future research and practice in hospitality? As we mentioned, industry leaders in both the hotel and restaurant industries report human capital challenges recruiting, retaining, motivating, training, and developing the workforce—as the problems that "keep them awake at night" (Enz, 2001, 2004). While short-term recruiting and turnover issues may ebb and flow with economic conditions, the general problem of acquiring, developing, and retaining a high-quality workforce that is ready to meet one's future organizational needs is a problem that does not go away in any financial environment. Obviously, industry leaders recognize the serious challenge of understanding the investment return from what is typically viewed in the short term, as a very significant P and L expense. Coupled with this contradiction is the fact that, as any GM, financial planner, and/or hotel controller would advise, critical strategic decisions must be based on sound financial analysis—even of an organizations crucial intangible assets, such as investments in its employees.

Yet, at the same time, it is not difficult to realize in a service-oriented business, how crucial employees, especially frontline employees, are to an organization's success. Building from the concepts discussed by Mr. Longstreet, as well as the recognition that frontline employees are the ones carrying out a firm's strategic initiatives (Liao & Chuang, 2004; Skaggs & Youndt, 2004), we offer that both hospitality-based HR research and practice should begin with these employees. In other words, when determining the return on investing in and developing a hospitality firm's human capital, both research and practice should begin with the crucial customer-employee interaction.

More specifically, hotel or restaurant guests have many "touch points" or encounters with employees, as throughout their visit, they consume the product for sale. We suggest focusing on the data embedded in these encounters. Through guest comments systems and client feedback programs, most organizations already know which aspects of their service delivery system are working and which are candidates for improvement. We argue that these data represent points for future investigation from not a service quality perspective but perhaps in a more nuanced manner, from an HR, operational one. For example, what are specific employees specifically doing to manage the service encounter in ways that the most valuable, repeat customers appreciate? Or alternatively, how do these employees create efficiencies around the work they do that are reflected in high-volume exchanges with satisfied customers and clients? How are efforts such as these reflected in additional revenues or cost savings to the hotel or restaurant? Questions such as these help identify what employees are doing right—all the critical and significant things they do to help build and sustain a successful business.

With an understanding of data such as these in hand, HR decision makers can then assess not only what outstanding performance "looks like" in terms of bottom-line impact, but they can also determine the routines that outstanding employees enact, that can be embedded in the organization's processes. At a broader level, decision makers can conduct similar analyses examining the impact of past and current investments in various employee groups (reflected in the wages and benefits required to obtain them) on the hotel's current revenue stream and even profitability. Ideas such as these are aimed at applying financial modeling to understand the impact of human capital at both the individual and group level.

The same type of analyses would be useful to analyze the impact of different HR investments. For example, with information from loyal guests in hand, decision makers could examine ways criteria provided by these guests are embedded into performance appraisal systems, are used to identify training standards, and/or are shaped into hiring guidelines for evaluating job candidates. In this way, HR policies and practices become aligned to not only support and develop these high-performing employees but also to encourage others to adopt similar behaviors. Practices such as these would help create and sustain a high-performing work culture. In addition, the success of HR initiatives could be calibrated against how well employees are succeeding in managing those crucial customer linkages that help generate repeat business.

The linkages between customer and employees represent one leverage point from which our ideas could be directly applied to hospitality research and practice. Connecting employee performance to what matters most to a hotel or restaurant, along with the ability to accurately strengthen and measure this connection through HR processes, metrics, and analytics pose exciting opportunities for decision makers. The impact of human capital and HR initiatives is obviously critical to a service-oriented business. Understanding the financial depth of this impact likely represents an occasion for hotel and restaurant organizations to differentiate themselves in significant ways from their competition. We look forward to future hospitality research and practice that focuses on this opportunity and offers a greater understanding of the strategic value of HR.

CONCLUSION

In this chapter, we sought to not only review key issues in SHRM but also to identify areas of opportunity and leverage. Our review made salient for us that research has a long way to go in providing definitive answers for hospitality practitioners. Yet, it is clearly moving in that direction. Applying metrics and adopting a financial decision-making framework will help all decision makers—whether HR or operational—examine the return associated with human capital investments and HR initiatives. The key lies in understanding ways these assets appreciate in value to the organization. The business leaders who can best leverage their human capital will be the ones contributing to the long-term strength of their companies.

REFERENCES

1. Amit, R., & Shoemaker, P. J. (1993). Strategic assets and organizational rent. Strategic Management Journal, 14, 33-46.
2. Argote, L., & Ingram, P. (2000). Knowledge transfer: A basis for competitive advantage in firms. Organizational Behavior and Human Decision Processes, 82, 150-169.
3. Barney, J. B. (1991). Firm resources and sustained competitive advantage. Journal of Management, 17, 99-120.

4. Barney, J. B., & Wright, P. M. (1998). On becoming a strategic partner: The role of human resources in gaining competitive advantage. Human Resource Management, 37, 31-46.

5. Bartels, R. (1965). Development of marketing thought: A brief history. In G. Swartz (Ed.), Science in marketing. New York: Wiley.

6. Becker, B., & Gerhart, B. (1996). The impact of human resource management on organizational performance: Progress and prospects. Academy of Management Journal, 39, 779-801

7. Becker, B., & Huselid, M. A. (1998). High performance work systems and firm performance: A synthesis of research and managerial implications. Research in Personnel and Human Resource Management, 76, 53-101.

8. Becker, B., Huselid, M. A., Pickus, P. S., & Spratt, M. F. (1997). HR as a source of shareholder value: Research and recommendations. Human Resource Management, 36, 39-47.

9. Becker, G. S. (1962). Investment in human capital: A theoretical analysis. The Journal of Political Economy, 70(5), 9-49.

10. Bontis, N. (1998). Intellectual capital: An exploratory study that develops measures and models. Management Decision, 36, 63-76.

11. Bontis, N., Crossan, M., & Hullard, J. (2002). Managing an organizational learning system by aligning stocks and flows. Journal of Management Studies, 39, 437-469.

12. Boudreau, J. W (1991). Utility analysis for decision in human resource management. In M. D. Dunnette & L. M. Hough (Eds.), Handbook of industrial and organizational psychology (2nd ed.) (Vol. 2, pp. 621-745). Palo Alto, CA: Consulting Psychologists Press

13. Boudreau, J. W, & Berger, C. J. (1985). Decision-theoretic utility analysis applied to employee separations and acquisitions [Monograph]. Journal of Applied Psychology, 70, 581-612.

14. Boudreau, J. W, & Ramstad, P. M. (1997). Measuring intellectual capital: Learning from financial history. Human Resource Management, 36, 343-356.

15. Boudreau, J. W, & Ramstad, P. M. (2003). Strategic HRM measurement in the 21st century: From justifying HR to strategic talent leadership. In M. Goldsmith, R. P. Gandossy, & M. S. Efron (Eds.), MM in the 21st Century (pp. 79-90). New York: Wiley.

16. Boudreau, J. W., & Ramstad, P. M. (2005). Talentship, talent segmentation and sustainability: A new HR decision science paradigm for a new strategy definition. Human Resource Management, 44, 129- 136.

17. Boudreau, J. W, Sturman, M. C, & Judge, T. A. (1994). Utility analysis: What are the black boxes and do they affect decisions? In N. Anderson 8c P. Herriot (Eds.), Assessment and selection in organizations (pp. 77-96). New York: Wiley.

18. Bouillon, M. L., Doran, B. M., 8c Orazem, P. F. (1995-1996). Human capital investment effects on firm returns. Journal of Applied Business Research, 12, 30-41.

19. Bowen, D. E. (1986). Managing customers as human resources in service organization. Human Resource Management, 25, 371-383.

20. Brockbank, W (1999). If HR were really strategically proactive: Present and future directions in HR's contribution to competitive advantage. Human Resource Management, 38, 337-352.

21. Carmeli, A., 8c Tishler, A. (2004). The relationship between intangible organizational elements and firm performance. Strategic Management Journal, 25, 1257-1278.

22. Coff, R. W. (1997). Human assets and management dilemmas: Coping with hazards on the road to resource-based theory. Academy of Management Review, 22, 374-402.

23. Delaney, J. T., & Huselid, M. A. (1996). The impact of human resource management practices on perceptions of organizational performance. Academy of Management Journal, 39, 949-969.

24. Duffy, J. (2000). Measuring customer capital. Strategy & Leadership, 28, 10-13.

25. Engstrom, T. E. J., Westnes, P., & Westnes, S. F. (2003). Evaluating intellectual capital in the hotel industry. Journal of Intellectual Capital, 4, 287-303.

26. Enz, C. A. (2001). What keeps you up at night?: Key issues of concern for lodging managers. Cornell Hotel and Restaurant Administration Quarterly, 4, 38-45.

27. Enz, C. A. (2004). Issues of concern for restaurant owners and managers. Cornell Hotel and Restaurant Administration Quarterly, 45(4), 315-332.

28. Gupta, S., & Lehmann, D. R. (2003). Cutomers are assets. Journal of Interactive Marketing, 17, 9-24.

29. Gutek, B. A., & Walsh, T. (2000). The brave new service strategy. New York: Amacom.

30. Hall, R. (1992). The strategic analysis of intangible resources. Strategic Management Journal, 13, 135- 144.

31. Hammonds, K. H. (2005). Why we hate HR. Fast Company, 97, 41-47.

32. Harrison, J. S., & Enz, C. A. (2005). Hospitality strategic management. Hoboken, NJ: Wiley.

33. Hitt, M. A., Bierman, L., Shimizu, K., & Kochhar, R. (2001). Direct and moderating effects of human capital on strategy and performance in professional service firms: A resource-based perspective. Academy of Management Journal, 44, 13-28

34. Huselid, M. A. (1995). The impact of human resource management practices on turnover, productivity and corporate financial performance. Academy of Management Journal, 38, 635-672.

35. Huselid, M. A., & Becker, B. E. (1995). High performance work systems and organizational performance. Paper presented at the 1995 Academy of Management Conference, Vancouver, BC

36. Huselid, M. A., Jackson, S. E., & Schuler, R. S. (1997). Technical and strategic human resource management effectiveness as determinants of firm performance. Academy of Management Journal, 40, 171-188

37. Jackson, S. E., Schuler, R. S., & Rivero, J. C. (1989). Organizational characteristics as predictors of personnel practices. Personnel Psychology, 42, 727-786.

38. Judge, T. A., Thoresen, C. J., Bono, J. E., & Patton, G. K. (2001). The job-satisfaction-job performance relationship: A qualitative and quantitative review. Psychological Bulletin, 127, 376-407.

39. Kannan, G., & Akhilesh, K. B. (2002). Human capital knowledge value added. Journal of Intellectual Capital, 3, 167-179.

40. Kundo, S. C, 8c Vora, J. A. (2004). Creating a talented workforce for delivering service quality. Human Resource Planning, 27, 40-51.

41. Lado, A. A., & Wilson, M. C. (1994). Human resource systems and sustained competitive advantage: A competency-based perspective. Academy of Management Review, 19, 699-727.

42. Latham, G. P., & Whyte, G. (1994). The futility of utility analysis. Personnel Psychology, 47, 31-46.

43. Lawler, E. E., Levenson, A., & Boudreau, J. W (2004). HR metrics and analytics: Use and impact. HR. Human Resource Planning, 27(4), 27-35.

44. Lawler, E. E., & Mohrman, S. A. (2003). HR as a strategic partner: What does it take to make it happen? Human Resource Planning, 26(5), 15-29.

45. Lepak, D. P., & Snell, S. A. (1999). The human resource architecture: Toward a theory of human capital allocation and development. Academy of Management Review, 24, 31-48.

46. Lepak, D. P., & Snell, S. A. (2002). Examining the human resource architecture: The relationships among human capital, employment and human resource configurations. Journal of Management, 28, 517-543.

47. Liao, H., & Chuang, A. (2004). A multilevel investigation of factors influencing employee service performance and customer outcomes. Academy of Management Journal, 47, 41-58.

48. Manz, C. C, 8c Sims, H. P. (1991). Superleadership: Beyond the myth of heroic leadership. Organizational Dynamics, 19, 18-35.

49. McGovern, T, 8c Panaro, J. (2004). The human side of customer relationship management. Benefits Quarterly, 20, 26-33.

50. Nelson, R. R., & Winter, S. G. (1982). The evolutionary theory of economic change. Cambridge, MA: Belknap Press.

51. Nonaka, I. (1994). A dynamic theory of knowledge creation. Organization Science, 5, 14-37.

52. Normann, R. (1984). Service management: Strategy and leadership in service businesses. Chichester, UK: Wiley.

53. Peteraf, M. (1993). The cornerstone of competitive advantage: A resource-based view. Strategic Management Journal, 14,179-191.

54. Reichheld, F. E (1996). The loyalty effect: The hidden force behind growth, profits and lasting value. Boston: Harvard Business School Press

55. Sasser, W. E., Schlesinger, L. A., & Heskett, J. L. (1997). The service profit chain. New York: Free Press.

56. Schneider, B., Ehrhart, M. G., Mayer, D. M., Saltz, J. L., & Niles-Jolly, K. (2005). Understanding organization- customer links in service settings. Academy of Management Journal, 48, 1017- 1032.

57. Sherer, P. D. (1995). Leveraging human assets in law firms: Human capital structures and organizational capabilities. Industrial and Labor Relations Review, 48, 671 -691.

58. Simons, T. (2002). The high cost of low trust. Harvard Business Review, 80(9), 18-19

59. Skaggs, B. C, & Youndt, M. (2004). Strategic positioning, human capital and performance in service organizations: A customer interaction approach. Strategic Management Journal, 25, 85-99

60. Sturman, M. C. (2000). Implications of utility analysis adjustments for estimates of human resource intervention value. Journal of Management, 26, 281-299.

61. Sturman, M. C. (2003). Introducing utility analysis for hospitality: A tool for quantifying the value of human resource interventions. Cornell Hotel and Restaurant Administration Quarterly, 44, 106-116.

62. Teece, D. J., Pisano, G., 8c Shuen, A. (1997). Dynamic capabilities in strategic management. Strategic Management Journal, 18, 509-534.

63. Ulrich, D. (1997). Human resource champions. Boston: Harvard Business School Press.

64. Ulrich, D., & Brockbank, W. (2005). Focusing on customers. HR Magazine, 50(6), 62-66.

65. Ulrich, D., & Smallwood, N. (2005). HR's new ROI: Return on Intangibles. Human Resource Management, 44, 137-142.

66. Way, S. A., Sturman, M. C, & Raab, C. (in press). What matters more? Contrasting the effects of job satisfaction and service climate on hotel F8cB managers' job performance. Cornell Hospitality Quarterly

67. Whyte, G., & Latham, G. P. (1997). The futility of utility analysis revisited: When even an expert fails. Personnel Psychology, 50, 601 -610.

68. Wright, P.M., Dunford, B. B., & Snell, S. A. (2001). Human resources and the resource based view of the firm. Journal of Management, 27, 701-721.

69. Wright, P. M., Smart, D. L., & McMahan, G. C. (1995). Matches between human resources and strategy among NCAA basketball teams. Academy of Management Journal, 38, 1052-1074.

70. Wright, P. M., & Snell, S. A. (1991). Toward an integrative view of Strategic Human Resource Management. Human Resource Management Review, 1, 203-235.

71. Wright, P. M., Snell, S. A., & Jacobson, P. (2004). Current approaches to HR strategies: Inside-out versus outside-in. Human Resource Planning, 27, 36-46.

72. Youndt, M. A., 8c Snell, S. A. (2004). Human resource configurations, intellectual capital and organizational performance. Journal of Managerial Issues, 16, 337-360.

73. Youndt, M. A., Snell, S. A., Dean, J. W, 8c Lepak, D. P. (1996). Human resource management, manufacturing strategy and firm performance. Academy of Management Journal, 39,836-866.

74. Zucker, L. G., Darby, M. R., 8c Brewer, M. B. (1998). Intellectual capital and the birth of U.S. biotechnology enterprises. The American Economic Review, 88(1), 290-306.

Chapter 7

Preparing Hospitality Organizations for Self-Service Technology

Joseph D. Lema

Hospitality Management, Drexel University, Philadelphia, PA

ABSTRACT

Self-service technology is rapidly changing the hospitality industry, providing new opportunities for the delivery of services and options for customers. Preparing to implement effective self-service technology delivery programs requires a workforce that can rapidly adapt to change. Understanding factors that influence employee readiness to engage in and support self-directed processes are an important consideration when implementing self-service technology. The results of the linear regression model in this study indicate that generalized self-efficacy and the self-directed learning readiness of employees in the hospitality industry are significantly related variables. While self-efficacy was the most highly correlated variable to the self-directed learning readiness of hospitality employees, future studies should consider other characteristics that may influence self-direction. As self-service technology continues to rapidly expand in all areas of the hospitality industry, opportunities and challenges exist for both employees and customers.

Keywords: Self-service technology, hospitality, self-directed learning

INTRODUCTION

Encompassing areas of food and beverage, lodging, and entertainment, the hospitality industry is one of the largest and fastest growing industries in the world, with an enormous amount of human capital investment in a diverse range of jobs. It is estimated that by the year 2014 the hospitality industry will employ more than 14,693 million workers in the United States (U.S. Department of Labor, 2005). With an industry that has one of the largest capital labor expenditures in the global economy, the need to focus on the foundation of the hospitality workforce includes an examination of employee development (Erdly & Chatterjee, 2003).

The rapid development of self-service technology is significantly influencing hospitality organizations by providing new opportunities and challenges for customers and employees. A *Time* magazine article reported on the popularity of self-service, with "U.S. customers spending $128 billion at self-service kiosks last year, an 80 percent jump from the year before, and by 2007 it could hit $1.3 trillion" (Kiviat, 2004, p. 101). To maximize learning capacity, self-directed learning readiness is a required strategy to consider with the self-service concept. Self-service allows users greater control over their experience, just as self-directed learning emphasizes the learner's personal control over his or her own learning experience (Long, 2000).

The skill level and adaptability of employees to initiate change in their jobs are factors in the competitiveness of a hospitality organization. With greater emphasis on productivity and accountability for individual performance, the responsibility for employees to rapidly adapt to change is moving from an organizational perspective to that of self-directedness. As the hospitality industry continues to consolidate, these organizations are experiencing mergers and acquisitions as rapidly as many individual employees change jobs and careers.

Technology in hospitality organizations has provided much of the success over past challenges with self-service strategies, although human capital investment is required to match the personal characteristics necessary to maximize the full potential of self-service benefits. A recent article in *The Economist* reported, "Self-service is now doing for the service sector what mass production once did for manufacturing, automating processes and significantly reducing costs" (You're Hired, 2004, p. 21). With self-service kiosks in hotels, restaurants, and airports, self-service options are becoming a part of everyday life. The explosion of online learning in education and business is another example; the result of rapid increases in technology. Although the capacity of self-service strategies is limited by the competency of the user, self-directed learning is at the center of the self-service concept. The purpose of this study was to examine how self-efficacy and selected demographic variables (position, gender, and ethnicity) relate to the self-directed learning readiness of employees in hospitality organizations.

Examining whether significant relationships exist among the variables of self-efficacy, demographics, and the self-directed learning readiness of employees in the hospitality industry will have an impact on the productivity and competitiveness of the hospitality industry workforce. Connecting these theoretical relationships in the dynamic and diverse environment of the hospitality industry will help link learning theories with emerging practical applications. Examining self-efficacy and demographic variables that influence the self-directed learning readiness of employees provides valuable insights into employee learning strategies and human resource development.

As hospitality organizations are required to remain competitive by implementing technologies that affect the learning environment of employees, the responsibility for self-directed learning has increased, along with technology. Characteristics that embrace self-directed learning and those that act as a barrier to self-directed learning need to be examined in order to determine the readiness of an employee to adopt self-directed learning (Long, 1991b). As reported by senior executives within the hospitality industry, some of the overarching issues for organizations include retention, education, training, and recruitment. In addition, positive solutions are needed to help transition employees and adopt strategies that initiate continuous change efficiently in targeted high-growth areas such as the hospitality industry (U.S. Department of Labor, 2005).Peter Senge (1990), known for his "learning organization" theory, reported in his widely referenced book, *The Fifth Discipline*, that the new learning organization comprises workers who can adapt to change with transformation through motivated self-directed learning and critical thinking.

While technology is fueling the growth of vast amounts of available information, employees are being challenged to apply relevant information in the context of their own work situation. As technology continues to rapidly advance in the hospitality industry and society, it is becoming extremely important to have a highly capable workforce. Developing a competitive workforce is important in a global economy, and it is essential in highly competitive areas such as the hospitality industry. While technology has fueled much of the growth in delivering new services and options that customers are demanding, a well-prepared and dynamic workforce is required to match the rapidly changing levels of technology.

Motivating employees to initiate change requires preparation and practice. Preparing an employee to be self-directed involves consideration of assumptions that follow self-directed learning methods. An understanding of the theoretical research of the variables that influence self-directed learning readiness will allow a hospitality organization to create effective, efficient programs and practices that maximize the talents of its employees. This unique workforce may have a much greater responsibility to meet the diverse and ever-demanding needs of customers.

Organizations and employees that promote self-directed learning readiness will help prepare their employees to participate in self-directed work teams and support a learning organizational strategy. One of the challenges of this study is to investigate the relationships that selected variables have on self-directed learning readiness. Developing a broader knowledge base of self-directed learning readiness with selected variables may not only benefit a people-based business that is highly concentrated in employing and serving people, but other researchers and practitioners may benefit from the unique social learning environment that the hospitality industry has to offer.

LITERATURE REVIEW

Self-directed learning has been a high-interest topic in the fields of business and education for more than a decade (Mezirow, 1985). Definitions of self-directed learning share a number of unique and similar perspectives. Although there are a number of unifying elements in defining self-directed learning, there is also a concerning ambiguity in precise definitions among the research (Oddi, 1987). Research into self-directed learning, according to Long (1991a), has developed over the past decade in both quantity and quality. Self-directed learning theory may have received greater attention than the practical aspect, which remains underdeveloped and has not received the same attention. Brookfield (1984) describes the process of self-directed learning as lacking a full appreciation for the impact of a skillful instructor and may fail to appreciate the social influence of subgroups in the surrounding community or environment. Part of the confusion with the self-directed learning term may be linked to learning as an internal change process and education as an external change process that facilitates internal change (Brockett & Hiemstra, 1991). Rather than moving away from self-directed learning, Brockett and Hiemstra advocate expanding the concept through continued development as a central theme and conceptual framework for adult learning.

Self-directedness is a characteristic of adult learning that is closely associated with self-directed learning and includes a level of decision making and personal control throughout the learning experience. Tough (1979) regards self-directed learning as a form of adult learning that includes the ability to plan and guide the learning process. Adults, according to Tough, have a desire to learn by drawing upon personal autonomy, self-worth, and acknowledgment of life experiences. Providing adults the opportunity to direct and plan their own learning builds self-directedness that supports a growing number of adult learning theories (Knowles, 1984/1998; Long, 1991b; Tough, 1979).

Self-Efficacy

Self-efficacy is the belief in one's own capability to initiate control over situations in an organized process (Bandura, 1994). Expectations of self-efficacy involve psychological procedures which when analyzed may be distinguished as two expectations of efficacy and outcome (Bandura, 1977).

"An efficacy expectation is the conviction that one can successfully execute the behavior required to produce the outcomes. An outcome expectation is defined as a person's estimate that a given behavior will lead to certain outcomes" (Bandura, 1977, p. 79). The differentiation between efficacy and outcome expectations is linked to the learner's belief that a course of action may produce certain outcomes, but the learner questions whether he or she can actually perform those actions.

Bandura (1977) argues that the strength of conviction of the learner's own belief in effectiveness may determine whether the learner will pursue changing or challenging situations. Learners may fear and avoid challenging situations when their belief is that they will not be able to handle the problem. Conversely, Bandura explains, learners may behave with confidence when they judge themselves to be capable of successfully handling situations that would have otherwise been threatening to them.

Self-efficacy theory is based on two types of expectations, mentioned earlier as efficacy expectations and outcome expectations, along with the characteristics, behavior, and behavioral outcomes of the person (Bandura, 1986). Efficacy expectation (self-efficacy) is the person's confidence in his or her ability to produce the behavior, while the outcome expectation results from the behavior based on a person's belief about the outcome. Self-efficacy may be a more accurate predictor of performance since outcome expectations are dependent upon self-efficacy (Bandura, 1986). Employees, for example, may be more motivated to perform behaviors that they believe will produce desired outcomes.

Using self-efficacy as a predictor, Bandura (1986) explains, is important in understanding how people function in terms of the choices they make (selection processes), effort (time and persistence), motivation (initiation), thought patterns (cognitive processes), and emotional reactions (affective processes) to various situations. The main sources of information that influence beliefs in self-efficacy include experience of mastery, observation, verbal persuasion, and physiological information (Bandura, 1986, 1997).

One of the most influential sources of information on self-efficacy is experience of mastery (Bandura, 1986). According to Bandura, success and failure attributes are important sources of information for developing self-efficacy. Successful experiences help enhance self-efficacy with a feeling of mastery and control, while repeated failure decreases self-efficacy over time (van der Bijl & Shortridge-Baggett, 2002). When a learner has developed a strong self-efficacy, explains van der Bijl and Shortridge-Baggett, the effect of one failure may not have much influence since the effects of failure follow a total pattern of experiences, although the timing of the moment in the learning process may vary in the power of the effect. If failure takes place in the early stages of the learning process, for instance, the greater will be its negative impact on self-efficacy (van der Bijl and Shortridge-Baggett, 2002).

Bandura (1986) describes a hierarchy in the sources of information for self-efficacy and categorizes them as direct and indirect sources of information. Experience of mastery, for example, is one of the most powerful sources of information, as a person experiences success or failure immediately based on direct information. The other information sources include observation of others, verbal persuasion, and physiological information based on indirect sources of information. Indirect sources of information may not be nearly as powerful in terms of information for self-efficacy as the cognitive process associated with critically reflective patterns of direct earlier experiences. Other sources that influence self-efficacy include personality traits (Strecher, DeVellis, Becker, & Rosenstock, 1986) such as self-esteem, locus of control, self-confidence, and hardiness (Coppel, 1980), and environmental factors such as expectations and support of others (Bandura, 1986).

In a further theoretical analysis of sources that influence self-efficacy, Gist and Mitchell (1992) suggest that experiences of mastery, observation, verbal persuasion, and physiological information contribute through a variety of internal and external information cues. Internal information cues relate to an individual's knowledge or skills and the person's effectiveness in using these skills through various strategies. An individual's self-efficacy can be determined by an internal assessment (adequate, inferior, or superior) of abilities when performing at various task levels. Judgments about expected performance when engaged in a task can be influenced by mood, health, or degree of arousal, whether positive (excited) or negative (fearful). External information cues relate to the characteristics of the task itself, such as complexity, number of components, parts, sequence, uncertainty, and steps. The resources and interdependence required to successfully complete the task can also influence the estimated level of self-efficacy.

Examinations of self-efficacy, Bandura (1997) suggests, often require assessments that an individual makes in terms of the variability in influencing determinants, previously described as experience of mastery, observation, verbal persuasion, physiological information, and others. The level of variability may provide sources of information that range from low to high, immediate or over longer periods, stable or unpredictable. Acquiring knowledge, for instance, is one factor that may have an immediate effect on an individual, whereas other factors such as ability may change after longer periods of time. Bandura also argues that immediate variability in a factor may result in greater perceived control over those factors that are relatively stable and require longer periods of time.

Bandura (1986) emphasizes that one important element to consider with self-efficacy is the perception of control. Some factors involve personal control (e.g., effort), while other factors are controlled by someone else (e.g., facilitator). The perception that the causes of performance are uncontrollable may result in lower levels of variability, resistance to change, and a lower level of self-efficacy. Bandura claims that analysis and understanding of the individual and task is necessary to enhance self-efficacy.

According to Bandura (1994), self-efficacy involves the belief that people have in their personal capabilities the ability establish personal standards. Since personal standards may be modified by the environment or demographic characteristics (position, gender, ethnicity), beliefs in personal capabilities may influence possible discrepancies between capabilities and self-generated standards. Based on his social cognitive theory, Bandura's reciprocal process of self-efficacy has a primary objective of enhancing learning skills and self-directedness in individuals (Bandura, 1994; Kitson, Lekan, & Guglielmino, 1995).

Self-efficacy theories have created a framework for understanding elements related to self-directed learning.Brockett and Hiemstra (1991), for example, developed a two-component model referred to as the Personal Responsibility Orientation (PRO) that supports personal responsibility and individual ownership of the learner's thoughts and actions or learner self-direction. The other component consists of self-directed learning that emphasizes the relationship between the learner and facilitator. The PRO model suggests that self-efficacy is central to understanding self-direction in regards to employee learning. The model also suggests that employees are capable of taking a proactive approach to learning and, when given the opportunity to be self-directed, there is the potential to maximize benefits for both the employee and organization.

Another model based on the situational nature of the learner and facilitator was developed by Grow (1991) called the Staged Self-Directed Learning Model (SSDL). This model assumes that the self-directedness of the learner is based on situational processes. Grow explains that learners progress through stages of self-direction that may either increase or decrease depending upon the situational circumstances. Furthermore, Grow argues, depending on the facilitator's approach, learning may be supported or hindered in the process. The SSDL model may help to indicate whether a facilitator's style aligns with the learner's self-directed learning readiness.

METHODOLOGY

The convenience sample consisted of employees who work in hospitality organizations, with approximately 216 employees participating. Data collection occurred at three participating hospitality organizations during April 2006. The participating organizations offer a diverse workforce, with food and beverage, lodging, and entertainment operations representing significant areas of the hospitality organization. A facilitator administered the survey to the employees who volunteered for the study in the participating organizations' business facilities. Completion time for the survey was 30 minutes. Participants were presented with an informed consent form before they participated in the study which clearly stated the voluntary nature of participation, the ability to withdraw from the survey at any time, and confidentiality of the participants' identities.

A survey integrating the Oddi Continuing Learning Inventory (OCLI) and Generalized Self-Efficacy Scale (GSE) was presented to the participants as a five-page instrument consisting of 49 questions. The OCLI and GSE instruments uniquely complement each other in their generalizability to provide a stable comprehensive indicator of relationships rather than measuring a narrowly defined activity that could be the result of a brief occurrence. Rather than measuring a specific task, the OCLI and GSE measure overall job-related activities. The GSE instrument was used to assess self-beliefs of personal capabilities of the employee (Jerusalem & Schwarzer, 1993). The OCLI instrument, developed as a doctoral dissertation by Oddi (1984), was administered to assess employees' levels of self-directed learning readiness. The relationships among GSE scores and OCLI scores were examined.

The OCLI survey is one of the most widely reliable and validated instruments used for the measurement of readiness for self-directed learning (Brockett & Hiemstra, 1991). The OCLI survey measures the level of self-directed learning readiness of adults. With a reported Cronbach's alpha of .88 and retest reliability of $r = .89$, the OCLI is an adequately reliable instrument for this study (Oddi, 1984).

Validation of the OCLI instrument was conducted by Oddi, Ellis, and Altman-Roberson (1990) to examine the relationship of the survey constructs and behavioral characteristics that indicate self-directed learning readiness. Three theories were developed to describe the affective, motivational, and cognitive attributes of the self-directed learner. The proactive drive versus reactive drive, commitment to learning versus apathy to learning, and cognitive openness versus defensiveness were reported by Oddi et al. (1990) to be the three constructs that emerged. Factor analysis reported by Oddi (1984) indicated that OCLI items contained self-confidence, autonomous learning, and learning with the participation of others. When items were loaded on a general factor analysis, reading avidity and self-regulation emerged as subsidiary factors. No factor was related to cognitive openness in the analysis, since scores failed to correlate with the adult intelligence factor. When scores failed to correlate with adult intelligence, discriminate validity was provided. The two other dimensions that Oddi describes as reading avidity and ability to be self-regulating positively correlated with self-confidence, participation, and endurance. These results indicate that the total OCLI score can be used to provide a reliable and valid measure for the construct of self-directed learning readiness.

The generalizability of the OCLI, detailed in a follow-up study by Six (1987), reported that factor analysis across different populations suggested that the factors identified by Oddi (1984) in the development of the OCLI instrument were not unique to the sample. The factor analysis indicated that the factors derived by Oddi did not break up under different study conditions to form new factors and, as a result, remained stable across different studies (Six, 1987). Validation of the factor match, Six argued, demonstrates the generality of the instrument across different populations. Respondents circled an answer from a 7-point Likert scale ranging from 1 (*strongly agree*) to 7 (*strongly disagree*) to best describe their behavior. The total self-directed learning readiness score from the survey was used in the statistical procedures as recommended in the literature (Brockett & Hiemstra, 1991; Oddi, 1984).

The GSE consists of a scale designed to measure the generalized self-efficacy or the employees' belief in their ability to perform their job. Jerusalem and Schwarzer developed the GSE in 1980, and it has been used with thousands of participants in 27 language versions throughout the world (Jerusalem & Schwarzer, 1993). The stability of the GSE has been reported in a number of longitudinal studies along with validation of the instrument in similar occupational and educational environments (Jerusalem & Schwarzer, 1993; Pasveer, 1998; Schwarzer, BaBler, Kwiatek, Schroder, & Zhang, 1997; Schwarzer & Born, 1997;Schwarzer, Mueller, & Greenglass, 1999; Schwarzer & Schroder, 1997). The GSE reports a Cronbach's alpha of .75, with a retest reliability (after 1 year) of $r = .67$ and a stability coefficient (after 2 years) of $r = .75$ (Jerusalem & Schwarzer, 1993). The participant was required to circle the correct response from a 4-point scale ranging from 1 (*not at all true*) to 4 (*exactly true*). The overall score of the instrument was calculated by totaling the response score to the appropriate questions. Scoring of the instrument was in accordance with the guidelines provided by the authors (Jerusalem & Schwarzer, 1993).

FINDINGS

Descriptive data for the total sample include position, gender, and ethnicity. The sample consisted of individuals working as supervisors (34%, $n = 71$) and nonsupervisors (66%, $n = 141$). In addition, the sample contained 52% females ($n = 111$) and 48% males ($n = 101$). Ethnicity included 14% African American ($n = 30$), 2% American Indian ($n = 5$), 17% Asian ($n = 35$), 45% Caucasian ($n = 96$), 13% Hispanic ($n = 27$), and 9% Pacific Islander ($n = 19$). Due to missing responses on the OCLI and GSE, four participants were eliminated from the study. Three participants inaccurately completed responses on the OCLI scale and one participant failed to complete responses on the GSE scale. Consequently the OCLI and GSE scales could not be sufficiently scored for these participants. A total of 212 participants provided an adequate sample size for the number of predictors used in the stepwise multiple linear regression model.

The OCLI variable consisted of a 7-point scale, with a lower number (24 being the lowest possible score) indicating less self-directed learning readiness and a higher number (168 being the highest possible score) representing greater self-directed learning readiness. Scores for the OCLI ranged from a low score of 51 to a high score of 154 with a mean score of 107. The range of OCLI scores in this research were consistent with other research in the area of self-directed learning readiness. The mean OCLI score in this study (107), however, was lower than the OCLI mean score (128) in a study on practicing nurses (Oddi, 1987).

The GSE is measured on a 4-point scale, with a lower number indicating less generalized self-efficacy and a higher number representing greater generalized self-efficacy. Scores for the GSE ranged from a low score of 12 to a high score of 40 with a mean score of 30. Industry experience was measured in months of hospitality industry work experience. Position was effect coded into the classifications of Supervisor and Nonsupervisor. Gender was effect coded into the categories of Male and Female. Finally, ethnicity was effect coded into the classifications of African American, American Indian, Asian, Caucasian, Hispanic, and Pacific Islander.

Overall Stepwise Multiple Linear Regression Model

A correlation matrix is presented in Table 1 to report the relationships between the OCLI and GSE, along with selected demographic variables. The GSE variable reported the strongest correlation with OCLI scores (r = .840) when compared to position, gender, and ethnicity. Position was negatively correlated with OCLI scores, indicating that supervisors had higher OCLI scores and nonsupervisors had lower OCLI scores. All variables reported a statistically significant correlation with OCLI scores at or below the .05 alpha level.

Table 1. Pearson Correlation Matrix (N = 212)

	OCLI	GSE	Position	Gender	Ethnicity
OCLI	—				
GSE	.840*	—			
Position	−.405*	−.414	—		
Gender	.310*	.089	.024	—	
Ethnicity	.129*	.006	−.105	.082	—

Note. *$p < .05$, one-tailed.

Considering the number of predictors and exploratory nature of the study, stepwise multiple linear regression was selected for the statistical analysis. Only significant predictors remain in the stepwise multiple linear regression model by the retesting of the predictor variables at each step (Mertler & Vannatta, 2005). The final stepwise multiple linear regression model consisted of four predictor variables, including generalized self-efficacy, position, gender, and ethnicity.

Data inspection did not locate any outliers, therefore no cases were deleted from the analysis. Evaluations of linearity, Kolmogorov-Smirnov tests of normality, homoscedasticity, and multicollinearity showed that the assumptions were within the range of tolerance. Four of the variables—generalized self-efficacy, position, gender, and ethnicity—entered into the overall model, $R^2 = .81$, $R^2_{adj} = .81$, $F(4, 206) = 176.48$, $p = .001$. The final stepwise multiple linear regression model accounted for 81.1% of the variance in OCLI scores.

Hypotheses Testing

Hypothesis 1: There is a significant relationship between self-directed learning readiness and self-efficacy of employees in the hospitality industry.

The results of the stepwise multiple linear regression model indicated that the self-efficacy variable makes a statistically significant contribution to the self-directed learning readiness of employees in the hospitality industry. The strength of the correlation ($r = .840$) between self-efficacy and self-directed learning readiness indicated that higher self-efficacy scores are associated with higher self-directed learning readiness scores. Furthermore, self-efficacy reported the strongest correlation among all of the tested variables to the self-directed learning readiness of employees.

Hypothesis 2: There is a significant relationship between self-directed learning readiness and selected demographic variables (position, gender, and ethnicity) of employees in the hospitality industry.

The stepwise multiple linear regression model reported that position had a statistically significant relationship to self-directed learning readiness of employees in the hospitality industry. The strength of the relationship was ($r = -.405$), indicating that supervisors have higher self-directed learning readiness scores than nonsupervisors. Position was the most strongly correlated variable to self-directed learning readiness of the demographic variables. Gender also reported a statistically significant relationship to the self-directed learning readiness of employees in the hospitality industry and entered into the stepwise multiple linear regression model. The strength of the relationship ($r = .310$) indicated that females had higher self-directed learning scores that males. Finally, ethnicity entered into the stepwise multiple linear regression model and had a statistically significant relationship to the self-directed learning readiness of employees in the hospitality industry. The strength of the relationship ($r = .129$), although weak, indicated that Asians had higher self-directed learning readiness scores.

DISCUSSION

Self-Efficacy

The findings of this study provided both expected and unexpected results in view of previous studies. Self-efficacy, as expected, significantly correlated with the self-directed learning readiness of employees in the study, yet the strength of the relationship compared to the other variables was surprisingly higher. Self-efficacy, as described by Bandura (1994), refers to the belief in one's own capability to initiate control over situations in an organized process. The strong, significant relationship between self-efficacy and self-directed learning readiness supports Bandura's (1997) notion that learners will pursue challenging situations when they have a belief that their capabilities to handle situations will produce positive outcomes. An employee's motivation to participate in self-directed learning activities comes from the belief that they are capable of succeeding in handling those particular situations.

The strength of the relationship between self-efficacy and self-directed learning readiness was discovered to be stronger than personal characteristics and demographic variables. In examining the four dynamics Bandura (1977) describes as performance accomplishments, experience, verbal persuasion, and emotional arousal, the importance of building positive experiences is also evident in the significance of the strength in the relationship between self-efficacy and the self-directed learning readiness of employees in this study. Since experience of mastery is one of the most important sources of information for determining levels of self-efficacy, positive experiences with self-directed learning opportunities may reciprocally indicate a greater readiness to engage in self-directed activities and higher levels of self-efficacy. Lower levels of self-efficacy imply lower levels of self-directed learning readiness based on the results of this study. The quality of the experiences, however, may not be determined from the findings of this research.

Demographic Variables

Position was found to be significantly correlated to the self-directed learning readiness of employees in this study. In addition, position entered into the final stepwise multiple linear regression model. Individuals in supervisory positions showed higher levels of self-directed learning readiness than those in nonsupervisory positions. Although the significant results of this variable are consistent with the findings of another study

(Roberts, 1986) involving a Hong Kong telephone company, other researchers have indicated mixed results in examining position relative to self-directed learning readiness (Brockett & Hiemstra, 1991). Oddi (1987) also reports that further examination of the variable "position" is recommended. Supervisor positions in the hospitality industry typically require greater leadership and critical decision-making responsibilities than nonsupervisors. The results of this study imply that supervisors have a greater self-directed learning readiness than nonsupervisors and, although statistically significant, the correlation is not nearly as strong as the self-efficacy variable.

Gender was significantly correlated with self-directed learning readiness. Furthermore, gender entered into the final stepwise multiple linear regression model. Coinciding with a previous study by Guglielmino and Guglielmino (as cited inBrockett & Hiemstra, 1991), females reported higher overall levels of self-directed learning readiness scores than males. Although their study reported a significant relationship between gender and self-directed learning readiness scores, the difference was narrow. The results of this study similarly reported a narrow correlational significance between self-directed learning readiness scores relative to the three other variables of interest, including self-efficacy, position, and ethnicity.

The results of ethnicity indicated a significant relationship with self-directed learning readiness scores. In addition, ethnicity entered into the final stepwise multiple linear regression model, indicating possible predictive capabilities in determining self-directed learning readiness for employees. Ethnicity has shown inconclusive results in a number of self-directed learning readiness studies in another field, nursing, as reported by Oddi (1987). The correlation between ethnicity and self-directed learning readiness is relatively weak in comparison to the other variables that were tested in this study.

CONCLUSION

The role of self-efficacy in relation to self-directed learning readiness will require careful consideration. An increase in the level of self-directed learning readiness of employees will need to coincide with strategies that enhance self-efficacy. AsBandura (1977) argues, experience of mastery is one of the most significant information sources relative to self-efficacy. In view of Bandura's theory, providing focused facilitation to complement self-directed processes may help to provide positive and successful experiences with self-service technologies and increased self-efficacy.

Bandura's (1986) self-efficacy theory also emphasizes that environmental factors can impact levels of self-efficacy. The social element of the hospitality industry provides unique situations where concentrations of employees and customers interact in dynamic environments. Levels of support may change in an instant for employees and customers, and social support may be instrumental in enhancing levels of self-efficacy. Organizational cultures that create an environment in which hospitality among employees and customers is an essential priority will help to positively influence self-efficacy and self-directed learning readiness strategies.

Organizations that are able to incorporate self-directed learning concepts into their self-service processes may benefit by delivering successful programs to their employees and customers. With further understanding in the differences of self-directed learning readiness, measurement, and benchmarking procedures among participants, hospitality organizations have an opportunity to gain competitive advantages. Recognizing self-directed learning as a dynamic process that varies among different groups of individuals based on their unique characteristics may help provide successful self-service programs for hospitality organizations.

Employee self-efficacy should be considered when implementing self-directed learning processes. Examining the self-directed learning readiness of employees will help to determine at what level employees are able to successfully engage in self-directed processes. The importance of building strong levels of self-efficacy relative to the self-directed learning readiness of employees is evident in the results of this study.

Self-efficacy may be easily enhanced in situations that offer immediate personal benefits to the learner. An employee who, for example, is a novice user of technology may be highly motivated to pursue a learning opportunity that will provide an immediate impact on his or her life, such as having unlimited access to self-service benefit options. An organization that can facilitate personal rewarding experiences for their employees also have an opportunity to create positive experiences that may increase self-efficacy and advance technological skill levels. One of the concerns thatHu, Nelson, Braunlich, & Hsieh (2003) explain in their research on technology-related training is that participants need to be more self-motivated in training activities in order to use the technological capabilities to the fullest extent. Providing learning opportunities through activities that have an immediate personal interest and impact on employees may be one possible motivational strategy. Furthermore, in view of rapid technological developments, offering employees incentives to purchase personal computers for their homes may provide other opportunities for employees to gain experience with self-service applications. Providing employees with the opportunity to gain self-service technology experience can begin with initiatives that are of personal interest to employees, such as self-service benefits enrollment, payroll transactions, and other personnel-related activities.

In moving beyond the misconceptions and misunderstandings of self-directed learning, Brocket et al. (2000) propose that by identifying new lines of inquiry into self-directed learning readiness, opportunities exist to fully expand the potential of self-directed learning. Oddi (1987) argues that opportunities exist to move beyond self-directed learning as a self-instructional process to examine self-directed learning in terms of cognitive, motivational, and affective characteristics and personalities of self-directed learners. Houle (1961) suggests, for example, the essence of self-directed learning is the inquiring mind that approaches life with openness to discovery. Houle argues that outstanding continuing learners have this attribute of personality (inquiry) to initiate learning. Although Oddi advocates that various modes of learning should not fail to be recognized, linking personality characteristics to self-directed learners offers substantial benefits to understanding the self-directed learning readiness of learners. Self-efficacy, being the predominantly significant variable relative to the self-directed learning readiness of employees as examined in this study, similarly supports Oddi's attempt to identify significant relationships related to self-directed learning processes.

Since the total OCLI and GSE scores are the most highly recommended and valid reporting statistics, factor analysis within the scales will not be used in further interpretation of the results due to a less stable level of reliability. Furthermore, another limitation of this study is in regards to the GSE instrument as a measure of generalized self-efficacy and not a measure of self-efficacy of a particular task, therefore limiting the analysis to generalized results. The results from this study did not attempt to provide cause-and-effect relationships among the variables. While offering suggestions for future research investigations to refine self-directed learning, careful consideration should be given to the operational definition of variables. The complexity of examining aspects related to self-directed learning may appear to be narrow in some circumstances, yet broad in others. Inquiry into self-directed learning, however, should continue to experiment with variables that will provide significant relationships and predictive capabilities with self-directed learning instruments to measure participants' levels of readiness in both educational and occupational environments.

Opportunities exist to further develop instruments that measure self-directed learning readiness. Both Oddi and Guglielmino have significantly contributed to advancing self-directed learning theories with the development of their self-directed learning readiness instruments. As technology continues to rapidly shape society and the hospitality industry, development of new instruments that build on the framework of existing self-directed learning readiness instruments will help to provide greater understanding of emerging self-directed learning processes.

Self-directed learning research should continue to benefit the hospitality industry as self-service technologies become part of daily operations. Hospitality organizations that are able to gain competitive advantages from self-service processes will need support in developing strategies that can provide the best experiences for their employees and customers.

REFERENCES

1. Bandura, A. 1977. *Social learning theory*, Englewood, NJ: Prentice Hall.
2. Bandura, A. 1986. *Social foundations of thought and action: A social cognitive theory*, Englewood, NJ: Prentice Hall.
3. Bandura, A. 1994. "Self-efficacy". Retrieved July 17, 2006, from http://www.des.emory.edu/mfp/BanEncy.html
4. Bandura, A. 1997. *Self-efficacy: The exercise of control*, New York: Freeman.
5. Brockett, R. G. and Hiemstra, R. 1991. *Self-direction in adult learning: Perspectives on theory, research, and practice*, New York: Routledge.
6. Brockett, R. G., Stockdale, S. L., Fogerson, D. L., Cox, B. F., Canipe, J. B., Chuprina, L. A., Donaghy, R. C. and Chadwell, N. E.2000, February. *Two decades of literature on self-directed learning: A content analysis*, (ERIC Document Reproduction Service No. ED449348) Boynton Beach, FL: Paper presented at the 14th International Self-Directed Learning Symposium.
7. **7.** Brookfield, S. D. 1984. Self-directed adult learning: A critical paradigm. *Adult Education Quarterly*, 36: 226–234.
8. Coppel, D. B. 1980. Relationship of perceived social support and self-efficacy to major and minor stresses (Doctoral dissertation, University of Washington, 1980). *Dissertation Abstracts International*, 41: 2751
9. Erdly, M. and Chatterjee, A. 2003, October. "The case for self-service hospitality". In *Hospitality Upgrade* Retrieved July 6, 2006, from http://www.hotelonline.com/News/PR2003_4th/Oct03_SelfServiceTech.html
10. Gist, M. E. and Mitchell, T. R. 1992. Self-efficacy: A theoretical analysis of its determinants and malleability. *Academy of Management Review*, 17: 183–206.
11. Grow, G. 1991. "The staged self-directed learning model". In *Self-directed learning: Consensus & conflict*, Edited by: Long, H. B. and Associates. 199–226. Norman: Oklahoma Research Center for Continuing Professional and Higher Education, University of Oklahoma.

12. Houle, C. O. 1961. *The inquiring mind*, Madison: University of Wisconsin Press.

13. Hu, B. A., Nelson, D. C., Braunlich, C. G. and Hsieh, Y. 2003. The future of Internet-based training in the lodging industry.*Journal of Human Resources in Hospitality & Tourism*, 2: 93–113.

14. Jerusalem, M. and Schwarzer, R. 1993. *The generalized self-efficacy scale* Retrieved July 16, 2006, from http://userpage.fu-berlin.de/~health/engscal.htm

15. Kitson, D. L., Lekan, D. F. and Guglielmino, P. J. 1995. "Self-directed learning readiness personality correlates". In *New dimensions in self-directed learning*, Edited by: Long, H. B. and Associates. 39–48. Norman: Public Managers Center, University of Oklahoma.

16. Kiviat, B. 2004, September. May you help you?. *Time*, 164(10): 101–102. Retrieved July 13, 2006, from EBSCOhost database,http://search.ebscohost.com/

17. Knowles, M. S. 1984. *The adult learner: A neglected species* , 3rd, Houston, TX: Gulf Publishing.

18. Knowles, M. S. 1998. *Self-directed learning: A guide for learners and teachers*, New York: Cambridge.

19. Long, H. B. 1991a. "Analysis of abstracts". In *Self-directed learning dissertation abstracts: 1966–1991*, Edited by: Long, H. B. and Redding, T. R. 5–14. Norman: Oklahoma Research Center for Continuing Professional and Higher Education of the University of Oklahoma.

20. Long, H. B. 1991b. *Self-directed learning: Consensus and conflict*, Norman: Oklahoma Research Center for Continuing Professional and Higher Education, University of Oklahoma.

21. Long, H. B. 2000. "The elusive concept of self-directed learning". In *Self-directed learning and the information age* Retrieved July 13, 2006, from EBSCOhost database, http://search.ebscohost.com/

22. Mertler, C. A. and Vannatta, R. A. 2005. *Advanced and multivariate statistical methods: Practical application and interpretation* , 3rd, Glendale, CA: Pyrczak Publishing.

23. Mezirow, J. 1985. "A critical theory of self-directed learning". In *Self-directed learning: From theory to practice*, (Report No. 25)San Francisco: New Directions for Continuing Education. (ERIC Document Reproduction Service No. EJ313257)

24. Oddi, L. F. 1984. Development of an instrument to measure self-directed learning (Doctoral dissertation, Northern Illinois University, 1984). *Dissertation Abstracts International*, 46: 01

25. Oddi, L. F. 1987. Perspectives on self-directed learning. *Adult Education Quarterly*, 38: 21–31.

26. Oddi, L. F., Ellis, A. J. and Altman-Roberson, J. E. 1990. Construct validation of the ODDI continuing learning inventory. *Adult Education Quarterly*, 40: 139–145.

27. Pasveer, K. A. 1998. Self-trust: Definition and measurement (Doctoral dissertation, University of Calgary, 1998). *Dissertation Abstracts International*, 59: 5170

28. Roberts, D. G. 1986. A study of the use of the self-directed learning readiness scale as related to selected organizational variables (Doctoral Dissertation, George Washington University, 1986). *Dissertation Abstracts International*, 47: 1218A–1219A.

29. Schwarzer, R., BaBler, J., Kwiatek, P., Schroder, K. and Zhang, J. X. 1997. The assessment of optimistic self-beliefs: Comparison of the German, Spanish, and Chinese versions of the generalized self-efficacy scale. *Applied Psychology: An International Review*, 46: 69–88.

30. Schwarzer, R. and Born, A. 1997. Optimistic self-beliefs: Assessment of general perceived self-efficacy in 13 cultures. *Worm Psychology*, 3: 177–190. Retrieved July 15, 2006, from EBSCOhost database, http://search.ebscohost.com/

31. Schwarzer, R., Mueller, J. and Greenglass, E. 1999. Assessment of the perceived general self-efficacy on the internet: Data collection in cyberspace. *Anxiety, Stress, and Coping*, 12: 145–161.

32. Schwarzer, R. and Schroder, K. E. 1997. Effects of self-efficacy and social support on postsurgical recovery of heart patients. *Irish Journal of Psychology*, 18: 88–103. Retrieved July 15, 2006, from EBSCOhost database,http://search.ebscohost.com/

33. Senge, P. 1990. *The fifth discipline*, New York: Doubleday Dell.

34. Six, J. E. 1987. Measuring the performance of the Oddi continuing learning inventory (Doctoral dissertation, Syracuse University, 1987). *Dissertation Abstracts International*, 49: 701

35. Strecher, V., Devellis, B. M., Becker, M. H. and Rosenstock, I. M. 1986. The role of self-efficacy in achieving health behavior change. *Health Education Quarterly*, 13: 73–91.

36. Tough, A. M. 1979. *The adult's learning projects: A fresh approach to theory and practice in adult learning* , 2nd, Toronto: Ontario Institute for Studies in Education.

37. **37.** U.S. Department of Labor. 2005. *Employment & training administration* Retrieved July 15, 2007, fromhttp://www.doleta.gov/BRG/Indprof/HPWI.cfm

38. van der Bijl, J. J. and Shortridge-Baggett, L. M. 2002. "The theory and measurement of the self-efficacy construct". In *Self-efficacy in nursing: Research and measurement perspectives*, Edited by: Lenz, E. R. and Shortridge-Baggett, L. M. 9–27. New York: Springer.

39. You're Hired. 2004. *Economist*, 372: 21–23. Retrieved July 9, 2007, from EBSCOhost database, http://search.ebscohost.com

Chapter 8

Recruitment and Selection Process and Employee Competence Outcome: An Important Area for Future Research

Mohammed Y. A. Alsabbah, Hazril Izwar Ibrahim

School of Management, Universiti Sains Malaysia, Penang, 11800, Malaysia

ABSTRACT

The literature treated the employee workplace competence term as a dependent variable when it was treated as one of human resource management outcomes. Particularly, more famous literature realizes the competence variable after training and development of new recruits, few studies have measured the competencies variable as one of the recruitment and selection process outcome. And a few researchers have judge worker's competency with regard to recruitment and selection. However, these competencies were heavily considered during the recruitment and selection criteria under the competence based recruitment and selection literature. Accordingly, the study will illustrate how the workplace competence was considered by previous studies, as well, the proposed conceptual framework will justify why the workplace competence is considered in evaluating the recruitment and selection outcome.

Keywords: Recruitment, Selection, Workplace Competence

1. INTRODUCTION

Hiring in the firm generally consists of two sets of activities. One set involves recruitment of applicants, while the second set involves screening and selection from among these applicants (Holzer, 1987). Past literature treated recruitment and selection as a combined (Shen & Edwards, 2004; Carless, 2007). However, if the applicant pool is inappropriate for the vacancy in the recruitment process, the effectiveness of the selection process becomes limited. Selection is the last phase of the recruitment activity (Giovanni, et al, 1995) and the terms "recruitment and selection" are generally considered to be fully interrelated processes (Hung, 2006).

A significant amount of literature has shown that a sound recruiting exercise yields high quality employees through a good selection method and can better predict future job performance (Gable, Hollon, & Dangello, 1992; Heraty & Morley, 1998; Mak, 1995; Murphy & Shiarella, 1997; Cascio, 1998). In addition, competency can effectively leverage individual team and organizational performance (Carroll & McCrackin, 1998). Furthermore from the researcher's perspective, because the unstable environment, as well the environmental pressure, the organizations must attain premium recruitment and selection process to attain competence and professional staff (Promís, 2008; Brown, 2007). It is also true in terms of success of work performance and perceived quality (Sangeetha, 2010), and the creation of a true competitive advantage for organizations (van Birgelen, Wetzels, & van Dolen, 2008). Many firms, therefore, spend millions of dollars in their efforts to recruit and select the best employee (Blackman, 2006).

The literature outlines several scientific terms of the outcome of recruitment and selection processes. For the recruitment and selection, previous studies often focused on the employee turnover(e.g. Bishop, 2005), performance (e.g. Breaugh, 1981; Schmidt & Hunter 1998), and quality terms (e.g. Williams, Labig, & Stone, 1993; Breaugh, et al. 2003). Little attention was considered for the employee workplace competence in the recruitment and selection outcome. However, these competencies were heavily considered during the recruitment and selection criteria under the competence based recruitment and selection literature (e,g. Grigoryev, 2006; Reio Jr & Sutton 2006; Snyder, Rupp &Thornton 2006; Keep and James, 2010.).

2. THE COMPETENCE DEFINITIONS

The literature discussed a wide range of definitions for competence. Spencer and Spencer (1993) defined a competency as "an underlying characteristic of an individual that is causally related to . . . superior performance in a job or situation" (p .9). Also competencies are attributes, specific constellation of an individual's characteristics such as knowledge, skills, motives, traits, behavior, aspects of one's self-image or social role, or attitudes that gives someone the potential for effective task performance (Vorgelegt von, 2008). It is also a range of capabilities that enable some people to meet the desired behavior or a range of work demands more effectively than others (Kurz and Bartram, 2002). As well Carroll &

McCrackin (1998) definition include "competencies are knowledge, skills, characterize excellent performance within a specific context" (p. 46).

Additionally, Promís (2008) integrated between quality and competence terms, since he utilized the terms as synenum, while conducting the "qualified" people to indicate candidates have the highest degree of the competencies (Promís,2008, p 28).

A number of studies consider competence and competency to be similar (Sanghi, 2004). It was also demonstrated the definitions of competencies was conducted synonymously. Cumming, et al., (2009) illustrates that such as a plethora of terms is in common usage, a number of which are used synonymously and interchangeably, and the meaning of skills is utilized to indicate numerous terms. Examples include skills, competence, attribute, quality, ability, capacity and capability. As well, Reidy (2004) utilized the term of skill or quality terms as synonymous. Along with the notion of competence, this definition has been used in many themes to describe levels of skills and knowledge as applied to concrete work situations (Sicilia, García-Barriocanal, & Alcalde, 2005; Vorgelegt Von, 2008; Andrews and Higson, 2008; Richens, 1999). Some authors (e. g. Goodstein and Davidson, 1998) uses both terms synonymously and equal competencies are the knowledge, skills and abilities that are needed for a particular task or job (Ree, Carretta, & Steindl, 2001; von Vorgelegt, 2008). Moreover, Evers & Rush (1996) and Williams (2005) were considering competence is utilized to be synonymous with skill.

Accordingly, drawing from the definition adopted by Cumming, et al., (2009), Tymon (2011), De La Harpe, Radloff and Wyber (2000), Jonson (2011), Moore, Cheng, & Dainty, (2002), ; "competence" as defined for the purpose of this research a concept that describes the behavioral prerequisites for job performance and organizational results, indicated by skills attribute, character, quality, ability, capacity and capability. The study will consider the definition as a competence concept that describes the behavioral prerequisites for job performance at workplace and organizational results, indicating skills, character, attribute, quality, ability, capacity and capability. As according to the objective of this study, the study will take into consideration the illustration of a variety of competencies definitions that are used synonymously.

3. THE COMPETENCE VARIABLE

The literature treated the employee workplace competence as a dependent variable when it was treated as one of human resource management outcomes (e,g., Ey, 2006). Specially, more famous literature realizes the competence variable after training and development of new recruits. This literature was considered the competencies theme for the reason that it was defined the employee training and development as a process to change the knowledge, skills and behaviors for the employees (e,g,. Van Dyk, Nel, Loedolff and Haasbroek, 1997; Cronje, Hugo, Neuland and van Reenen, 1994; Avrabos, 2005).

In general, for measuring the employee workplace competencies, some studies focus on the competencies itself (e. g, Reio Jr & Sutton, 2006; Lievens, Sanchez, Bartram, & Brown, 2010). However, uncommonly studies have measured the competencies variable as one of the recruitment or selection process outcome. And few researchers judge worker's competency with regard to recruitment or selection, through evaluating the new workers competency (Bishop, 1983). Such competency considered for assessing recruitment and selection which includes aspects of the candidate, such as skills qualities (Vorgelegt von, 2008).

However, in relation to recruitment and selection, the literature focused on the competencies is rare and scatter. A few studies were conducted this relation, but in deficient manner, such as Brown (2007) study which confirms the correlation with employee competence as the outcome from the recruitment process. However, it utilized skills based on individual performance term to describe consideration of the board competence. Additionally, MacKenzie, Ployhart, Weekley, & Ehlers (2010) and Cabrera & Nguyen (2001) investigated only the selection method in relation to (KSA) theme from the employee competence.

However, in relation to one or two parts of recruitment and selection processes, excess of the studies was measured the competencies by conducting the skills within the job performance process (e.g. Lee & Phan, 2000). Also, Bandow (2004) behaves the skills measure for new hired under performance expectations theme. As well as Rodriguez, Patel, Bright, Gregory, & Gowing, (2002) use of competencies model to identify high-performing people. Moreover, Hunt (2007) states that the staffing assessments have been created to measure the attributes that influence performance. As well, Vathanophas (2007) consider the competency is an individual characteristic that can be measured to differentiate significantly between superior and average performers. That consideration was proven by literature, because of these studies classified the competencies theme to indicate the skills and abilities that an employee must have to predict or achieve successful performance (Spencer & Spencer, 1993). The assessment of competencies process were also measured for several objectives such as performance and training (Anitha Thenmozhi, 2011).

Furthermore, Modern Healthcare (2006) uses employee performance to measure employee competence in the workplace. Lee & Phan (2000) propose processes for the evaluation of board members. Such activities include performance measurements of board effectiveness in terms of competency attainment. Hunter & Hunter (1984) performed a Meta-analysis of the studies assessing several selection methods for predicting job performance for entry-level jobs. A study by, Schmidt & Hunter (1998) highlight the use of job performance in evaluating employee characteristic and skills.

Additionally, supplementary literature such as (Williams, Labig, & Stone, 1993; Breaugh, Greising, Taggart, & Chen, 2003; Keung & Pine, 2000) reviewed the competencies as employee quality, since it contains the employee skills and abilities. Moreover, in relation to recruitment and selection, other literature treated the competencies by reflecting the employee effectiveness. Forbes and Milliken (1999) listed two criteria of board effectiveness, to contain the board competencies (Lee & Phan, 2000). Further studies also were deemed the competencies to measure successful workplace term (Reio Jr & Sutton, 2006).

The literature considers the workplace competence as dependent variable but does not connect with recruitment and selection as independent variable. Such thought perceived by Brumm, Hanneman, & Mickelson (2006) study when validating 14 measurable competencies necessary and sufficient to measure curricular program outcomes.

Furthermore, for competencies measure, and within the Employment Service (ES), Rodriguez et al., (2002) constructs a pilot competency-based performance evaluation system using competency benchmarks. Competencies were selected as critical elements in the performance system, based on the distinguishing value (i.e., which competencies were rated the highest for distinguishing superior performers). These competency ratings are combined with ratings on the achievement of results to provide a complete picture of an individual's performance.

Studies by MacKenzie, Ployhart, Weekley, & Ehlers (2010) and Witt & Burke (2002) studies conducted an incompleted image of recruitment and selection, when using only the selection methods in relation to employee skills. Also, MacKenzie, et al (2010); Cabrera & Nguyen (2001); Oostrom (2010); MacKenzie Jr, Ployhart, Weekley, & Ehlers (2009); Stokes, Hogan, & Snell (1993); Weekley, Ployhart, & Harold (2004) performed the selection method, not in relation to competence theme, but related to (KSA) theme such as cognitive ability and personality.

In addition, to indistinguishable relation with employee competence, this relation also investigated under the performance expectation theme, for example Bandow (2004) just conducted the skills are the employers expected from the new hired subsequent of the hiring process, but the study did not conduct an empirical examination of the relation between recruitment and selection and competencies.

Moreover, this particular relationship has also been investigated by previous literature in an ambiguous and indefinite nature. Such that nature was performed the competence or skills as a parts of employee quality which considered by (Williams, Labig, & Stone, 1993; Breaugh, et al. 2003; Keung & Pine, 2000; Scholarios & Lockyer, 1999). Otherwise studies researched these competencies under "Predict Occupational Success term" to study selection methods outcome (Barrett & Depinet, 1991; McClelland, 1998). Also, the relationship between the two variables of recruitment and selection and competence variable was ambiguous when it was studied under dimensions of job performance in performance appraisal rating, such that concern was presented by (Taylor & Schmidt's, 1983; Zottoli & Wanous, 2000; Breaugh, 1981; Barrick & Zimmerman, 2009; Williams, et al., 1993) in the recruitment process. Schmidt & Hunter (1998) Murphy & Shiarella (1997) Borman, White & Dorseys (1995) Gomez (1985) and Schmidt, Hunter & Outerbridge (1986) also invastigaed the selection process. Sangeetha K (2010) and Fallon (2009) also measured recruitment and selection by linking it with employee effectiveness. Some other combination between skill term and performance term includes"performance-related skills" term to measure the selection method outcome (Damitz, Manzey, Kleinmann, & Severin, 2003)

3.1. Reasoning for Preferring Competence Variable

The current research conducted the employee competencies to determine the outcome of the recruitment and selection for attaining the employee skills. As well to determines if recruitment and selection decision was suitable or not (Sutherland & Wöcke, 2011). Also, the assessment of competencies provides management and staff with a common understanding of the skills and behaviors that are relevant to the organization. For that important concern, the organizations are turning into a pay for competencies strategy, since these organizations behavior strategic core competencies for success (Ledford & Heneman, 1999). As well, resent recruitment studies carry out that recently the companies have thought exclusively in terms of competence profile and not in terms of position success profile. However, they think more in terms of employee competence (Aurélie & Fallery, 2009).

It is necessary to highlight the fundamental values of competency, since today grown the use of competencies. It have also been used in many aspects of human resource management, ranging from employees recruitment and selection, development, and performance for strategic future planning in the organization (El-Goully & Wahba, 2011). However, the high competence employees are critical for high-performing organizations. Also, employeecompetencies were considered by literature as one factor that most contributed to the sustainability and , creation of organizational excellence (Vathanophas, 2007). Employee competencies were considered as a source for attaining competitive advantage in competitive circumstances for organizations (Sanghi, 2004). Also, employee competencies are needed in order to increase the business success (Zakaria, Zainal, & Nasurdin, 2011). The competencies are considered by the researchers as the future for both the employee and the institution

(McClelland, 1973; Rodriguez, Patel, Bright, Gregory, & Gowing, 2002), The use of competencies conducted to identify high competence for high-performing people, or outstanding employees, has gradually become widespread in human resource management (Boyatzis, 1982; Spencer & Spencer, 1993).

4. THE RELATIONSHIP BETWEEN VARIABLES

The literature considers the relationship between recruitment and selection and employee competence as a recipe for success. Such employer practices ultimately influence the employee job success (Gill, 2007, Vance & Foundation, 2006). Thus, the employers invest resources in recruitment and selection process because they expect the investment to enable them to hire the highest quality and competent workers (Bishop, Barron, & Hollenbeek, 1983). This is indicated by organizations successfully recruiting and selecting competent and quality employees (Howard, 1999; Bozionelos, 2005). Cascio, (1998) considers all phases of recruitment and selection should be with the appraisal of competence for job success.

An organization that uses a proper process to recruit and select skilled candidates will experience a higher level of job success (Gill, 2007; Vance & Foundation, 2006; Howard, 1999) and the employer who conducted an appropriate recruitment and selection process will enable the organization to receive employee quality and workers who embrace competencies (Bishop,et al., 1983; Bartram et al. 2002; McClelland, 1998; vorgelegt von, 2008; Breaugh et al, 2003; Ley & Albert, 2003;. Ey, 2006; Bishop, et al., 1983; Yng Ling, 2003; Keep and James, 2010)

Under Performance Expectations theme, Bandow (2004) examined the competencies for newly hired IT professionals in the US subsequent of the hiring process. From the pilot study, the author opines that Personal or individual as well as technical skills are the most skills should appear in the new hiring.

Fallon (2009) examines the relationship between the recruitment and selection process and the effectiveness of boards of health in USA, competencies were utilized to measure the effectiveness of the board, these competencies were: contextual, interpersonal, intellectual, political and strategic. The findings indicates that "appointers" as well as "the health commissioners" ratings (recruitment and selection) and board of health members' ratings (board competence for effectiveness) found no significant correlations.

According to Carless (2007) the recruitment and selection outcomes were to perceive the high quality of new recruits.

Furthermore, Brown (2007) illustrated the competencies definition when presented with recruitment domain, and developed a model to determine if using recommended recruitment resulted in more competent board members which led to the better board performance. Furthermore, the study considered the recruitment process under "Board Development" theme in relation to the employee competence.

Breaugh et al., (2008) considered the employee competencies and qualities as recruitment result, when they construct their recruitment models. As well as, Castilla (2005) and Chien & Chen (2008) identified the outcome of recruitment for the organization since the outcome determined the type of applicant's skills ability,and competencies need from the new hires. However, the competencies and quality of new recruits depends upon an organization's recruitment practice (Anderson and Shackleto, 1986).

Willie, Jayawardena, & Laver (2008) is convinced that the direct relationship between the recruitment process and employee qualities and competencies. However, it was identified the best approaches that should embrace success and attract high quality human resource talent within the Niagara region's hospitality industry.

Rynes & Boudreau (1986) revealed a significant relationship between recruiting practices, and perceived employee qualities. Positive relationships were considered between perceived effectiveness recruitment and new hires qualities and competencies. In this consideration, the study considers quality and professionalism of new hires such technical expertise, and early job performance. Also, directors of college recruiting for Fortune 1000 corporations were surveyed by Rynes & Boudreau, (2006) where they consider the relation of a broad set of college recruiting practices and college recruiting effectiveness in quality and competencies.

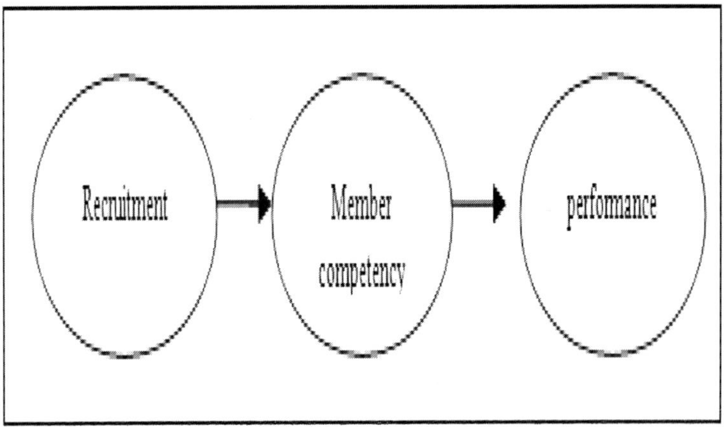

Figure 1. The relationship between recruitment, competency and performance, Brown (2007)

Moreover, Sangeetha K, (2010) conducted new-hire competencies for a construct "metrics" for measuring recruitment competence and effectiveness, metrics also help in the competencies measurement of the new candidate. This "metrics" can help in removing the inefficient candidates out of the system.

Table 1. Relation between Recruitment & Selection and Competency

Authors & Year	Country	Sample	Dependent variable	Dependent variable (the competencies)	Finding
Bishop, Barron, & Hollenbeek (1983)	USA	5,421 employers surveyed	Referral sources Employment (Agency, School Referral, Government Agency)	Knowledge, relevant work experience	-New hired employee was recruited by referrals will require less training because the competencies and skills they have at work. -Some recruitment sources typically yield better workers skills for performance than others -Employer referrals and new hires will require less training.
Barrick & Zimmerman, (2009)	USA	(N=119) from hired applicants	Recruitment sources and employee referral	Job knowledge, interpersonal skills, Communication skills, initiative, and punctuality.	-Applicants, who knew current employees by such referral such as employee referral; and number of friends and family, were higher skills in performance.
Bretz Jr & Judge, (1998)	USA	Rate the quality of the (N = 58) applicants resumes	Recruitment methods realistic job preview RJP	Work experience, and overall applicant competence	Applicant competence was not related to the weight placed on negative information
Gault, Leach & Duey (2010)	USA	185 unique employers of the undergraduate business interns	Internship recruitment methods	Eagerness to learn, Constructive, Demonstration of initiative and self-motivation, acceptance of criticism.	- Internships enhancing student competencies and career success -The majority of employers felt the internship contributed value to the intern in terms of future competencies for job performance.
Rynes, & Boudreau,(1986)	USA	145 Directors of college recruiting of Fortune 1000 corporations	Recruitment methods and communicate and advertisement	Critical skill areas in professional and technical expertise and quality	-Differentiate between recruiting sources and method and communication. -Positive relationships between perceived employee effectiveness in competencies for performance, and recruiter method, also with, recruiter selection criteria
Collins & Han (2004)	USA	99 organizations	Advertisement recruitment method	Applicable work experience skills that met the requirements of the position professional	Corporate advertising had direct effects on applicant competence and quality
Castilla (2005)	USA	334 of the employees hired	Personal contacts recruitment method	Work Experience Skills, Computer ,and Language Skills	The employee had got personal contacts appears superior skilled
Barrick & Zimmerman, (2009)	USA	(N=119) from hired applicants	Selection methods	Job knowledge and quantity of work. interpersonal skills, Communication skills, initiative, and punctuality.	Biodata check predicted skills for job performance rating.
Patterson, et al, (2005)	UK	46 doctors	Assessment Centre selection method	Professional integrity; and clinical expertise	Assessment Centre method can improves the employee competencies
Schmidt & Hunter (1998)	USA	meta-analytic of implications of 85 years of research in personnel selection	selection method	Conducting work job knowledge, competencies	The validity of 19 selection methods for predicting the competencies for ability to train and job performance, as well conscientiousness tests, work sample tests, job knowledge tests.
Borman, White & Dorsey's (1995)	USA	493 to 631 first-tour US Army soldiers	Cognitive ability test	Thinking competence, and Interpersonal competencies	Validity of cognitive ability test for considering competencies
Bertua, Anderson & Salgado, (2005)	UK	meta-analysis	Validity of tests of general mental ability (GMA)	Cognitive and professional competencies	Validity of tests for predicting competence for training success and performance

Furthermore, based on the basis of systemic approach recruitment and selection, Hunthausen, (2000) investigates recruitment and selection system validity of the US airline based on predictors task and contextual performance and multidimensional job performance. The author utilized supervisor competence evaluation to judgment incumbent job performance in the dimensions and skill measurements job performance, they were: written communication skills; oral communication skills; organizational ability; negotiation skills; quantitative skills; decision making ability; analytical ability\ problem; teamwork; employee development; flexibility; initiative, perseverance; leadership; and contentious improvement.

The evidence from Schmidt & Hunter (1998) study shows that different selection methods and combinations of methods have very different validities for predicting future employee competencies. Some methods, such as "interests and amount of education investigation", have very low validity. Others, such as graphology, have essentially no validity. While others, such as GMA tests and work sample measures have high validity.

Murphy & Shiarella (1997) argue that in order to estimate the validity of general cognitive ability test and personality tests, is to predict the competence for job performance, where performance is conceptualized as a composite of multiple performance measures such as organization citizenship behaviors competence. Furthermore, Hunter & Hunter (1984) conducted Meta-analysis to highlight the various selection methods for predicting the current employee characteristics for job performance, the study result shows the work sample test, with a mean validity of .54, is slightly better selection methods.

A Meta-Analytic Review conducted by Tett, Jackson, & Rothstein (1991) highlight the validity of personality measures of the selection methods as predictors employee characteristics for job performance, Also, Tett & Christiansen (2007) conducted the personality measures can contribute positively to employee characteristics such as Leadership. Furthermore, Reilly & Chao (1982) carry out the tests are valid for employee selection and employee characteristics conducted by supervisor rating.

Barrett & Depinet (1991) progress a meta-analysis to carry out the important of intelligence tests for predictive and measure the competence. As well, Hough & Oswald (2000) conducted the meta analysis to describe different selection methods such as cognitive ability test and verbal, numerical and spatial tests. However, the study considers a positive inter-correlation outcomes such as specific abilities.

In relation to the selection decision, and from the evidence provided when selection decision was not suitable, Sutherland & Wocke (2011) highlighted that the employee did not demonstrate expected skills, then the result of unexpected training time and expenditure to bridge those skills shortage and gap.

Judge & Ferris (1992) indicate that the selection decision resulted by an outflow selection decision. However, the study model was considered the inflow selection decision, resulted by the outflow selection decision, and mediated by through flow selection decision. Also, the employee competence and quality were reflected in the outflow decision. However, this decision was to regard the decision criteria.

Moreover, Wang, Li, & Hu (2009) validate the proposed Weighted Support Vector Machine for selection decision. However, for confirming the using of weighted support vector machine (WSVM), the study was investigated the employee competencies for job performance, because it gives the selection decision a feedback in their competencies.

Proposed framework

Recruitment and selection would be positively related to incumbent competence. However, based on the preceding discussion of the literature, the following is a proposed conceptual framework:

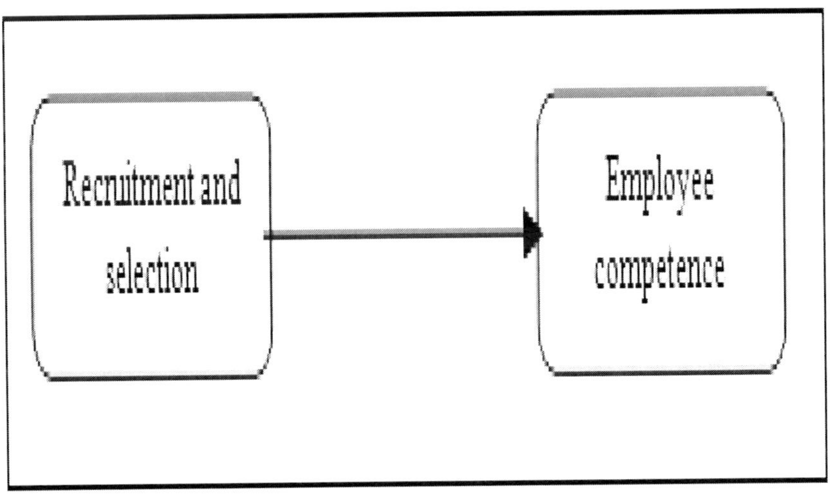

Figure 2. The conceptual framework of the relationship between recruitment & selection, and competency

5. CONCLUSIONS

In order to highlight the main contributions of the current study, the researcher argues that relatively little literature had conducted recruitment and selection, and seldom studies realize the relationship between recruitment and selection processes, and the employee competencies. Later studies can advance the contribution of the current study in order to contact empirical investigation for this proposed relationship.

REFERENCES

1. Andrews, J., & Higson, H.. Graduate employability,'soft skills' versus 'hard'business knowledge: A European study. Higher Education in Europe, 33(4), 411-422, 2008.
2. Aurélie, G., & Fallery, B. E-recruitment: new practices, new issues. An exploratory study. Human Resource Information Systems, 39-48, 2009.
3. Bandow, D. The shadow knows: Performance expectations of new hires in an IT organization, 2004.
4. Barrett, G. V., & Depinet, R. L. A reconsideration of testing for competence rather than for intelligence. American Psychologist, 46(10), 1012, 1991
5. Barrick, M. R., & Zimmerman, R. D. Hiring for retention and performance. Human Resource Management, 48(2), 183-206, 2009.
6. Bartram, D., Robertson, I. T., & Callinan, M. Introduction: A framework for examining organizational effectiveness. Organizational Effectiveness, 1-10, 2002
7. Bertua, C., Anderson, N., & Salgado, J. F. The predictive validity of cognitive ability tests: A UK meta-analysis. Journal of Occupational and Organizational psychology, 78(3), 387-409, 2005.
8. Bertua, C., Anderson, N., & Salgado, J. F. The predictive validity of cognitive ability tests: A UK meta-analysis. Journal of Occupational and Organizational psychology, 78(3), 387-409, 2005.
9. Bishop, J., Barron, & Hollenbeek. Recruiting Workers: How Recruitment Policies Affect the Flow of Applicants and Quality of New Workers, 1983.
10. Blackman, A. Graduating students' responses to recruitment advertisements. Journal of Business Communication, 43(4), 367, 2006.

11. Borman, W. C., White, L. A., & Dorsey, D. W.. Effects of ratee task performance and interpersonal factors on supervisor and peer performance ratings. Journal of Applied Psychology, 80(1), 168, 1995.

12. Boyatzis, R. E. The competent manager: A model for effective performance: John Wiley & Sons Inc, 1982.

13. Bozionelos, N. When the inferior candidate is offered the job: The selection interview as a political and power game. Human Relations, 58(12), 1605-1631, 2005.

14. Breaugh, J. A. Relationships between recruiting sources and employee performance, absenteeism, and work attitudes. Academy of Management Journal, 142-147, 1981.

15. Breaugh, J. A., Greising, L. A., Taggart, J. W., & Chen, H. The Relationship of Recruiting Sources and Pre-Hire Outcomes: Examination of Yield Ratios and Applicant Quality. Journal of Applied Social Psychology, 33(11), 2267-2287, 2003.

16. Breaugh, J. A., Macan, T. H., & Grambow, D. M. Employee recruitment: Current knowledge and directions for future research. International Review of Industrial and Organizational Psychology 2008, 45-82, 2008.

17. Breaugh, J. A., & Starke, M. Research on employee recruitment: So many studies, so many remaining questions. Journal of Management, 26(3), 405-434, 2000.

18. Bretz Jr, R. D., & Judge, T. A. Realistic job previews: A test of the adverse self-selection hypothesis. Journal of Applied Psychology, 83(2), 330, 1998.

19. Brown, W. A. Board development practices and competent board members: Implications for performance. Nonprofit Management and Leadership, 17(3), 301-317, 2007.

20. Brumm, T. J., Hanneman, L. F., & Mickelson, S. K. Assessing and Developing Program Outcomes through workplace competencies. International Journal of Engineering Education, 22(1), 123, 2006.

21. Cabrera, M. A. M. D., & Nguyen, N. T. Situational judgment tests: A review of practice and constructs assessed. International Journal of Selection and Assessment, 9(1-2), 103-113, 2001.

22. Carless, S. A. Graduate recruitment and selection in Australia. International Journal of Selection and Assessment, 15(2), 153-166, 2007.

23. Carroll, A., & McCrackin, J. The Competent Use of Competency-Based Strategies for Selection and Development. performance improvement quarterly, 11(3), 45-63, 1998.

24. Cascio, W. The employee recruitment and selection process Book, 1998.

25. Castilla, E. J. Social Networks and Employee Performance in a Call Center1. American Journal of Sociology, 110(5), 1243-1283, 2005.

26. Chien, C. F., & Chen, L. F. Data mining to improve personnel selection and enhance human capital: A case study in high-technology industry. Expert Systems with Applications, 34(1), 280-290, 2008.

27. Collins, C. J., & Han, J. Exploring applicant pool quantity and quality: The effects of early recruitment practice strategies, corporate advertising, and firm reputation. Personnel Psychology, 57(3), 685-717, 2004.

28. Crawford, L. Profiling the competent project manager, 2000.

29. Cumming, J., Kiley, M., Thomas, M., Hort, L., Pike, M., Evans, E., et al. Research Graduate Skills Project: Australian Learning and Teaching Council, 2009.

30. Damitz, M., Manzey, D., Kleinmann, M., & Severin, K. Assessment center for pilot selection: Construct and criterion validity and the impact of assessor type. Applied Psychology, 52(2), 193-212, 2003.

31. El-Goully, F., & Wahba, K. Proposing a Competence–Based Model for Recruiting Entry-level Directors in Egyptian Multinational Pharmaceutical Companies, 2011.

32. Evers, F. T., & Rush, J. C. The Bases of Competence. Management Learning, 27(3), 275-299, 1996.

33. Fallon, M. M. A quantitative study of the appointment process of local board of health members in Ohio and the relationship to board effectiveness. Bowling Green State University, 2009.

34. Gable, M., Hollon, C., & Dangello, F. Increasing the Utility of the Application Blank: Relationship Between Job Application information and subsequent performance and turnover of salespeople. The Journal of Personal Selling & Sales Management 12(3), 39, 1992 .

35. Gault, J., Leach, E., & Duey, M. Effects of business internships on job marketability: the employers' perspective. Education+ Training, 52(1), 76-88, 2010.

36. Gill, D. S. Employee selection and work engagement: do recruitment and selection practices influence work engagement? : ProQuest, 2007.

37. Gomez, J. J. The Evaluation of the Management Assessment Center. Final Report, 1985.

38. Goodstein, L. D., & Davidson, A. D. Hiring the" Right Stuff": Using Competency-Based Selection. Compensation and Benefits Management, 14, 1-10, 1998.

39. Henderson, F., Anderson, N., & Rick, S. Future competency profiling: validating and redesigning the ICL graduate assessment centre. Personnel Review, 24(3), 19-31, 1995.

40. Heraty, N., & Morley, M. In search of good fit: policy and practice in recruitment and selection in Ireland. Journal of Management Development, 17(9), 662-685, 1998.

41. Hough, L. M., & Oswald, F. L. Personnel selection: Looking toward the future--Remembering the past. Annual Review of Psychology, 51(1), 631-664, 2000.

42. Howard, J. L. Recruiting And Selecting Quality Employees: A Guide For Small Business And Entrepreneurs, 1999.

43. Hunt, S. T. Hiring success: the art and science of staffing assessment and employee selection: Pfeiffer, 2007.

44. Hunter, J. E., & Hunter, R. F. Validity and utility of alternative predictors of job performance. Psychological bulletin, 96(1), 72, 1984.

45. Hunthausen, J. M. Predictors of task and contextual performance: Frame-of-reference effects and applicant reaction effects on selection system validity. Portland State University, 2000.

46. Jattuso, M. L., & Sinar, E. F. (2003). Source Effects in Internet-based Screening Procedures. International Journal of Selection and Assessment, 11(2-3), 137-140.

47. Judge, T. A., & Ferris, G. R. The elusive criterion of fit in human resources staffing decisions. Human Resource Planning, 15, 47-47, 1992.

48. Keep, E., & James, S. Recruitment and Selection–the Great Neglected Topic. SKOPE Research Paper, 88, 2010.

49. Keung, S. W. C., & Pine, R. Changes in hotel industry recruitment as reflected by content analysis of newspaper advertisements. Asia Pacific Journal of Tourism Research, 5(2), 57-69, 2000.

50. Keunga;, S. W. C., & Pineb, R. Changes in hotel industry recruitment as reflected by content analysis of newspaper advertisements Asia Pacific Journal of Tourism Research, 5: (2,), 57 — 69, 2000.

51. Kurz, R., & Bartram, D. Competency and individual performance: Modelling the world of work. Organizational Effectiveness, Book, 227-255, 2002.

52. Le Deist, F. D., & Winterton, J. What is competence? Human Resource Development International, 8(1), 27-46, 2005.

53. Ledford, G., & Heneman, R. L. Pay for skills, knowledge and competencies. The compensation handbook: A state-of-the-art guide to compensation strategy and design, 143-156, 1999.

54. Lee, S. H., & Phan, P. Competencies of Directors in Global Firms: requirements for recruitment and evaluation. Corporate Governance: An International Review, 8(3), 204-214, 2000.

55. Ley, T., & Albert, D. Identifying employee competencies in dynamic work domains: methodological considerations and a case study. J. UCS, 9(12), 1500-1518, 2003.

56. Lievens, F., Sanchez, J. I., Bartram, D., & Brown, A. Lack of consensus among competency ratings of the same occupation: Noise or substance? Journal of Applied Psychology, 95(3), 562, 2010.

57. MacKenzie, J., Ployhart, Weekley, & Ehlers. Contextual Effects on SJT Responses An Examination of Construct Validity and Mean Differences Across Applicant and Incumbent Contexts. Human Performance, 23, 1–21, 2010.

58. MacKenzie Jr, W. I., Ployhart, R. E., Weekley, J. A., & Ehlers, C. Contextual effects on SJT responses: An examination of construct validity and mean differences across applicant and incumbent contexts. Human Performance, 23(1), 1-21, 2009.

59. Mak, C. Successful people selection in action. Health Manpower Management, 21(5), 12-16, 1995.

60. McClelland, D. C. Testing for competence rather than for" intelligence.". American Psychologist, 28(1), 1, 1973.

61. McClelland, D. C. Identifying competencies with behavioral-event interviews. Psychological Science, 9(5), 331-339, 1998.

62. Modern Healthcare. Ten Best Practices for Measuring the Effectiveness of Nonprofit Healthcare Boards [Copyright of Modern Healthcare is the property of Crain Communications Inc. (MI) and its content may not be copied or emailed to multiple sites or posted to a listserv without the copyright holder's express written permission. However, users may print, download, or email articles

for individual use. Source: Modern Healthcare, Dec: Bulletin of the National Center for Healthcare Leadership, Item Number: 2009919833], 2006 .

63. Moore, D. R., Cheng, M. I., & Dainty, A. R. J. Competence, competency and competencies: performance assessment in organisations. Work Study, 51(6), 314-319, 2002.

64. Murphy, K. R., & Shiarella, A. H. Implications of the multidimensional nature of job performance for the validity of selection tests: Multivariate frameworks for studying test validity. Personnel Psychology, 50(4), 823-854, 1997.

65. Oostrom, J. K. New Technology in Personnel Selection: The Validity and Acceptability of Multimedia Tests: Erasmus University Rotterdam, 2010.

66. Patterson, F., Ferguson, E., Norfolk, T., & Lane, P. A new selection system to recruit general practice registrars: preliminary findings from a validation study. BMJ, 330(7493), 711-714, 2005.

67. Promís, P. Are Employers Asking for the Right Competencies?: A Case for Emotional Intelligence. Library Leadership and Management, 22(1), 24-30, 2008.

68. Ree, M. J., Carretta, T. R., & Steindl, J. R. Cognitive ability. Handbook of industrial, work and organizational psychology, 1, 219-232, 2001.

69. Reilly, R. R., & Chao, G. T. VALIDITY AND FAIRNESS OF SOME ALTERNATIVE EMPLOYEE SELECTION PROCEDURES1. Personnel Psychology, 35(1), 1-62, 1982.

70. Reio Jr, T. G., & Sutton, F. C. Employer assessment of work-related competencies and workplace adaptation. Human Resource Development Quarterly, 17(3), 305-324, 2006.

71. Richens, G. P. Perceptions of Southern Nevada Employers Regarding the Importance of SCANS Workplace Basic Skills, 1999.

72. Rodriguez, D., Patel, R., Bright, A., Gregory, D., & Gowing, M. K. Developing competency models to promote integrated human resource practices. Human Resource Management, 41(3), 309-324, 2002.

73. Rynes, S. L., & Barber, A. E. Applicant attraction strategies: An organizational perspective. Academy of Management Review, 286-310, 1990.

74. Rynes, S. L., & Boudreau, J. W. College recruiting in large organizations: Practice, evaluation, and research implications. Personnel Psychology, 39(4), 729-757, 1986.

75. Sangeetha, K. Effective Recruitment: A Framework, 2010.

76. Sanghi, S. The handbook of competency mapping: understanding, designing and implementing competency models in organizations: Sage, 2004.

77. Sara l. Rynes, & boudreau, j. W. College recruiting in large organizations: practice, evaluation, and research implications. 39(4), 729-757, 1986.

78. Schlechty, P. C., & Vance, V. S. Recruitment, selection, and retention: The shape of the teaching force. The Elementary School Journal, 83(4), 469-487, 1983.

79. Schmidt, F. L., & Hunter, J. E. The validity and utility of selection methods in personnel psychology: Practical and theoretical implications of 85 years of research findings. Psychological bulletin, 124(2), 262, 1998.

80. Schmidt, F. L., Hunter, J. E., & Outerbridge, A. N. Impact of job experience and ability on job knowledge, work sample performance, and supervisory ratings of job performance. Journal of Applied Psychology, 71(3), 432, 1986.

81. Scholarios, D., & Lockyer, C. (1999). Recruiting and selecting professionals: Context, qualities and methods. International Journal of Selection and Assessment, 7(3), 142-156.

82. Sicilia, M. Á., García-Barriocanal, E., & Alcalde, R. Fuzzy Specializations and Aggregation Operator Design in Competence—Based Human Resource Selection. Soft computing: methodologies and applications, 219-230, 2005.

83. Spencer, M. Spencer, 1993. Competence at Work Models for Superior Performance: John Wiley & Sons, Inc., New York, 1993.

84. Stokes, G. S., Hogan, J. B., & Snell, A. F. Comparability of incumbent and applicant samples for the development of biodata keys: The influence of social desirability. Personnel Psychology, 46(4), 739-762, 1993.

85. Sutherland, & Wöcke. The symptoms of and consequences to selection errors in recruitment decisions S.Afr.J.Bus. Manage, ,42(4), 23-32, 2011.

86. Taylor, M. S., & Schmidt, D. W. A Process-Oriented Investigation Of Recruitment Source Effectiveness1. Personnel Psychology, 36(2), 343-354, 1983.

87. Tett, R. P., & Christiansen, N. D. Personality tests at the crossroads: A response to Morgeson, Campion, Dipboye, Hollenbeck, Murphy, and Schmitt (2007). Personnel Psychology, 60(4), 967-993, 2007.

88. Tett, R. P., Jackson, D. N., & Rothstein, M. Personality measures as predictors of job performance: a meta-analytic review. Personnel Psychology, 44(4), 703-742, 1991.

89. van Birgelen, M. J. H., Wetzels, M. G. M., & van Dolen, W. M. Effectiveness of corporate employment web sites: How content and form influence intentions to apply. International Journal of Manpower, 29(8), 731-751, 2008.

90. Vance, R. J., & Foundation, S. Employee engagement and commitment: A guide to understanding, measuring and increasing engagement in your organization: SHRM Foundation, 2006.

91. Vathanophas, V. Competency requirements for effective job performance in the Thai public sector. Contemporary Management Research, 3(1), 45-70, 2007.

92. vorgelegt. von. Decision making in personnel selection: A policy-capturing analysis in emerging German IT-firms book, 2008.

93. Weekley, J. A., Ployhart, R. E., & Harold, C. M. Personality and situational judgment tests across applicant and incumbent settings: An examination of validity, measurement, and subgroup differences. Human Performance, 17(4), 433-461, 2004.

94. Williams, C. The Discursive Construction of the" Competent" Learner-Worker: From Key Competencies to" Employability Skills". Studies in Continuing Education, 27(1), 17, 2005.

95. Williams, C. R., Labig, C. E., & Stone, T. H. Recruitment sources and posthire outcomes for job applicants and new hires: A test of two hypotheses. Journal of Applied Psychology, 78(2), 163, 1993.

96. Willie, P. A., Jayawardena, C., & Laver, B. Attracting and retaining quality human resources for Niagara's hospitality industry. International Journal of Contemporary Hospitality Management, 20(3), 293-302, 2008.

97. Witt, L., & Burke, L. Selecting high-performing information technology professionals. Journal of Organizational and End User Computing (JOEUC), 14(4), 37-50, 2002.

98. Yng Ling, Y. A conceptual model for selection of architects by project managers in Singapore. International journal of project management, 21(2), 135-144, 2003.

99. Zakaria, N., Zainal, S. R. M., & Nasurdin, A. M. Investigating The Role Of Human Resource Management Practices On The Performance Of Sme: A Conceptual Framework, 2011.

100. Zottoli, M. A., & Wanous, J. P. Recruitment source research: Current status and future directions. Human Resource Management Review, 10(4), 353-382, 2000.

Chapter 9

A Case Study into the Benefits of Management Training Programs: Impacts on Hotel Employee Turnover and Satisfaction Level

Youngsoo Choi & Duncan R. Dickson

Rosen College of Hospitality Management, University of Central Florida, Orlando, Florida, USA

ABSTRACT

Demonstrating the effectiveness of human resources (HR) programs has been a dilemma for HR professionals for years. This study addresses that problem and takes a step in closing the gap in the lodging industry. The researchers reviewed the effects of rigorous management training programs that focused on frontline leaders and their relationship with the employees who interact with guests. Data from a 2-year period (during and after the intervention) were analyzed based on employee-related factors, primarily turnover and employee satisfaction as measured by exit surveys. The researchers measured the impact of training over time. The results suggest this approach can be used by HR professionals to argue that management training programs lead to greater employee satisfaction and a reduction of employee turnover that offsets the cost of the training.

Keywords: Management training, turnover, employee satisfaction,

exit surveys

INTRODUCTION

Human Resources (HR) departments have been searching for proof that they were getting a significant return on investment (ROI) since the early 20th century when they were called "Personnel" departments. HR departments have been working to justify their expenditures for employee training to upper management and to accountants so they could continue designing the programs they intuitively believed were essential. Will the time and expense devoted to training increase productivity, profitability, employee satisfaction, and guest satisfaction enough to warrant the expenditure? It is obvious that many hospitality companies are not convinced training pays off. An American Society for Training and Development (ASTD) report indicated that only 8% of U.S. companies collect data and estimate their ROI for training (Gomez-Mejia, Balkin, & Cardy, 2007). Given this, is it any wonder that training departments are some of the first victims of cutbacks during economic downturns?

HR leaders have long struggled to be included in executive meetings. Finding a way to quantify the benefits of training programs could be a ticket to these important leadership sessions. Creating the image of HR as a value-adding department and not just a group to handle government regulations is a topic of concern in much of HR literature. One of the most frequently asked questions is, "Does HR pay for itself?" More and more of the traditional HR functions have been outsourced, including benefits administration, compliance training, risk management, labor, and in some instances, staffing. The message is clear in a business environment. Things can only be justified if they add to quarterly profits. If you cannot quantify a positive result, do not request funding.

The hospitality industry is a highly guest service–oriented business where encounters between employees and guests determine the success of the business. Consequently, hospitality researchers and business leaders have acknowledged the significance of hiring and retaining competent employees and have arguably considered the company's employees as the most valuable asset to hospitality firms. Therefore, training and retention quantifications are an important avenue to explore.

Even though there are numerous studies on Human Resource Management (HRM) practices, few studies have been conducted to empirically demonstrate and measure the effectiveness of certain HRM practices that applied to the lodging industry, which puts a high priority in reducing its employee turnover. The objective of this study is to contribute to the literature on HRM practices by describing a pilot study of a lodging company where new and revised management training programs have been implemented to its supervisory workforce over a period of time. The specific goals of the study are to:

1. Identify appropriate dependent variables (i.e., outcome measures) to explore the effectiveness of new management training programs;

2. Demonstrate the existence, and research significance of, a time lag that may be crucial to measuring the impact of HRM interventions; and

3. Measure the impacts of management training programs from two employee-related sources, one from primary data (employee responses to an exit interview survey) and the other from secondary data (employee turnover).

LITERATURE REVIEW

HRM practices, programs, and training have been developed and applied to the areas of recruiting and selection, training, benefits administration, and enhancement of employee perception or satisfaction over the years (Baumann, 2003;Lee, Heard, & Koh, 2000). The relationship between the various HRM practices and actual performance of the company has been examined in previous studies. These studies identified a list of dependent variables, such as employee turnover rate, employee satisfaction level, and company profit to measure the effectiveness of HRM practices (Batt & Valcour, 2003;Humber, 2003; Huselid, 1995; Kalleberg & Moody, 1994; Lee et al., 2000; Lloyd, 2000). For the purpose of this study, turnover rates and employee satisfaction levels have been chosen as the appropriate variables to measure the real benefit and effectiveness of the HRM practice in question: management training programs. Company profit was not considered because it may be affected by other significant factors, such as the economic situation, the development of new products, or a change in customers' preferences or tastes.

Employee Turnover

Employee turnover has been a major managerial issue in the hospitality industry, where the annual turnover rate is estimated to be as high as 300% (Hotel News Resource, 2000). Consequently, there are a number of studies on the culture of employee turnover in the lodging industry. Iverson and Deery (1997), in their study on the employees of five-star hotels in Australia, empirically showed that the turnover culture itself is the most important determinant of employees' intent to leave.Lloyd (2000) showed the existence of a significant disconnection between what companies believe causes employee turnover and why workers actually leave. Therefore, there is a dire need for managers to have a better and more accurate understanding of the actual causes of employee turnover.

It has been well documented that employee turnover involves a significant cost for the business. Turnover costs are categorized as separation costs, replacement costs, training costs, loss of productivity, and loss of knowledge (Pinkovitz, Moskal, & Green, 2002; Smith & Watkins, 1978; Tracey & Hinkin, 2006). A study estimated that replacing a non-management employee costs as much as 30% of the employee's annual salary, whereas replacing a manager incurs almost 50% of the manager's annual salary (Development Dimensions International, Inc., 2002).

Employee turnover has a direct impact on a company's productivity and financial performance. Harris, Tang, and Tseng (2002)used the Australian Business Longitudinal Survey data collected during 1995–1998 to try to quantify the impact of employee turnover on an organization's productivity. They also tried to derive the optimal turnover rate that maximizes a company's productivity and still allows for employee growth. In their study of multiple hotels in the United States,Simons and Hinkin (2001) provided an empirical test to show a strong association between high employee turnover and a decrease in hotel profits.

Despite the fact that there are a plethora of studies on employee turnover, relatively little research has been conducted to explore if management training programs have impact on reducing employee turnover.

Employee Satisfaction

Researchers in HR management have pointed out that departing employees can provide considerable insight into the problems they encountered while working for the company (Tracey & Hinkin, 2008; Woods & Macaulay, 1987). While the incumbent employees create the urgency for a company to act on the survey results, exit interviews enable a company to review the data and look for trends to better identify the fundamental problem areas in the workplace to enhance employee satisfaction levels and to reduce the turnover rate (Carvin, 2002).

Even though the significance of an exit interview has been acknowledged both in academia and in business, it is questionable whether companies have utilized actual employee exit interview data in order to make constructive improvements in their policies or practices. A study by Tracey and Hinkin (2006) stressed the value of exit interviews and demonstrated that there exits a negative correlation between the company's expenditure on employee exit interview process and total employee turnover costs. In order to best utilize departing employees' opinions, it is crucial to develop an effective structure of exit interview questionnaires. A literature review identifies the following categories as the most relevant for exit interviews: job, management or supervision, company policies, and salary and benefits (HR Solutions, 2003; Personnel Policy Service, 2003;Sweeney, 2002; Woods

& Macaulay, 1987). The data acquired from a well-designed exit interview can serve as a good measurement of HRM effectiveness.

Measuring the Benefit and Effectiveness of HRM Practices

The effectiveness of HRM practices in reducing employee turnover has been supported by a case study of New Castle Hotels. This case study identifies an open door policy, a peer review program, regular opinion surveys and an enhanced promotion policy as the main factors that reduced the New Castle Hotels' turnover rate from 66.5% in 1999 to 57% in 2000. These are considered nontraditional programs that are based on values such as fairness, consistency, caring, and ethics (Hotel News Resource, 2000).

Several studies have empirically tested the impact of HRM practices on reducing employee turnover. Lloyd (2000) stressed the importance of internal surveys, exit interviews, and an open-door policy to facilitate open communication between employees and management. Cho, Woods, Jang, and Erdem (2006) also emphasized the importance of HRM practices such as labor-management participation programs, incentive plans, and pre-employment testing to maintain a high retention rate among nonmanagerial employees. Humber (2003), in a case study concerning Sofitel, suggested the significance of implementing comprehensive recruiting and training programs in order to reduce the turnover rate of the company.

The quality of the employee selection system has been identified as one of the most important factors in keeping good workers and reducing the costly employee turnover (Development Dimensions International, 2002; Huselid, 1995; Poe, 2003). Enhancing the quality of managerial leadership is also considered a key to improving the turnover problem (Development Dimensions International, 2002; Hotel News Resource, 2000). Lee et al. (2000) showed that the economic situation matters when it comes to the effectiveness of HRM practices on employee turnover. They identified job analysis as the most effective practice during the boom years and succession planning as the most effective during recession periods.

Motivating current employees and improving their satisfaction has drawn the attention of many researchers (Allen, Shore, & Griffeth, 2003; Harris et al., 2002; Poe, 2003). Sturman and Trevor (2001) noted that many organizations collect the data on employee performance over time but rarely utilize the wealth of data collected to help predict and manage employee turnover. The study pointed out that training, job enrichment or job change, and reward systems are key practices to retaining quality employees.

Love (2004) also emphasized the significance of turnover costs, which include but are not limited to recruiting, training, and loss of knowledge capital, and suggested analyzing the turnover-related data in order for the company to predict future turnover. Chevalier, Siebert, and Viitanen (2003) used personnel data from a panel of 400 shops in a U.K. retail chain to investigate the effect of pay incentives on labor turnover and productivity. They showed that a flat hourly wage system without any reward for tenure or individual productivity led to the phenomenon of negative selection—only employees with no or little outside options remain with the firm—and consequently created lower productivity.

A literature review reveals gaps in the areas of study sample and design that need more attention. A majority of the study results are based on findings from surveys conducted among business professionals, mostly HR managers. Surveying only managers creates a one-sided look at the issues that senior leaders and financial officers would find suspicious. By broadening the study to those directly affected by the training programs, a more robust examination can be developed, especially when line-level employees are included. As Huselid (1995) criticized, existing literature contains a methodological weakness due to the issue of "simultaneity" or even reverse-causality between HRM practices (independent variables) and the dependent variable(s) of the studies. Simultaneity issues have developed because many empirical studies could not afford to wait for a period of time after specific HRM programs were put into action. Hence, the study results might be seriously tainted by low internal validity.

Methodology

Instead of relying on the feedback from managers or HR professionals in order to measure the impact of management training programs, this study investigated changes in employee-related factors, such as employees' turnover and satisfaction levels. While the managers and HR leaders' impressions of the programs' success are really immaterial, feedback from the actual line-level employees are believed to be a valid outcome measure of the intervention.

Study Site

The study was conducted with a lodging company that is located in the northeastern United States. The company operates two hotels. One is a historic, Colonial-style leisure and meeting hotel with just over 220 rooms. The other, located in the same city, is a full-service conference hotel with over 150 rooms when the study was conducted. The average number of employees in the company's HR database was about 700, of which about 75% were part-time, line-level employees.

As described before, a major initiative by this company was to develop and implement quality training programs for its managerial personnel. Through the analysis of outcomes, specifically employee turnover data and exit survey responses in reference to employee satisfaction, this pilot study suggests an approach that can be implemented by others in the industry to quantify the benefits of quality management training programs.

HRM Interventions: Enhancing the Quality and Rigor of Manager Training Programs

Several major training programs were developed and implemented within the company in 2002 with a specific focus on its managerial workforce. A four part training program was implemented for leaders. First, mandatory training for current and new managers was implemented company-wide in order to help them conduct the hiring and selection process more effectively. Training on applicant interviewing technique and on a Web-based application process was required for those involved in hiring. Second, comprehensive training programs to enhance the awareness of modern workplace challenges were developed so that managers became more capable of dealing with issues, such as sexual harassment, workforce diversity, and the employee discipline. Third, a training program was implemented to help managers recognize and utilize a type of scientific index that analyzes the new employees' job-related characteristics and their strengths and weaknesses. The purpose of this third initiative was for the managers to have a better understanding of the significance of matching new employees with the jobs that are more appropriate fit for their work-related characteristics. Finally, a standardized and rigorous employee performance appraisal system was established. Previously, each department has utilized its own appraisal system with different degrees of rigor and frequencies. A well-structured performance appraisal system based on actual compensable factors with high validity was applied to every department of the hotel after its managerial personnel were given the new appraisal training (Table 1).

In order to demonstrate the effectiveness of those new or revised management training programs, the dependent variables were chosen that measured how these interventions actually affected hotel line employees over time. The major rationale of this decision is that the benefit of managerial training programs needs to be quantified by measuring their impact on line employees rather than by asking perceptions of HR professionals or operation managers; improving the quality and rigor of management practices will help those managers conduct their HR responsibilities more effectively, which will enhance employee satisfaction and consequently reduce their turnover; and the real impact of these new initiatives, just like other policies and programs, will be realized after a certain period of time.

Table 1 Intervention and its Major Contents

Management training programs	Major contents of the training programs
Hiring/selection training	• Interview techniques • Legal concerns during the interviews • Effective utilization of a Web-based application process
Trainings to enhance awareness of modern workplace challenges among managers	• Sexual Harassment • Workforce diversity • Employee discipline
Technical training	• Effective utilization of a scientific index that analyzes the new employees' job-related characteristics and their strengths and weaknesses
Employee performance appraisal training	• Standardization of an employee performance appraisal systems across the departments

Employee Turnover Data

For this study, annual employee turnover rate was calculated by dividing total number of annual employee termination by average annual number of employees. The total number of annual employee terminations was acquired from the company's HR database, which included both resignations and discharges. The average annual number of employees was calculated by adding up all 12 monthly average numbers of employees and dividing it by 12.

The employee turnover rates for the years 2002 and 2003 were calculated. The employee turnover rate for 2002 represented the one when the key management training programs were first implemented. The turnover rate of 2003—1 year after the implementation of those programs—was calculated to see if there was a significant change in the company's turnover rate between 2002 and 2003 in order to allow for the impact of the new interventions to be realized in the workplace.

Exit Survey Data

After each employee left, an exit survey form was sent via regular mail along with a postage-prepaid return envelope. The total of 31 survey questions were developed based on a literature review of existing survey formats. Those items represent factors that are expected to measure employee satisfaction levels in regard to job/company, supervisor/manager, policies/practices, and pay/benefits. Respondents were asked to rate their satisfaction level on each item using a 5-point Likert scale ranging from 1 (very dissatisfied) to 5 (very satisfied). Two weeks after each termination, a survey packet was sent to the mailing address provided by the employee. The survey packet was not mailed to the former employees who were discharged to minimize a potential concern of outliers in the data set. A total of 506 (2002) and 290 (2003) exit surveys were sent out.

RESULTS AND DISCUSSIONS

The data analysis showed that the company achieved significant improvement in both reducing its employee turnover rate and enhancing employees' satisfaction level.

Result 1: Reduction in Employee Turnover Rates

During the period of study, this company was able to reduce the total number of annual employee turnover from 660 in 2002 to 383 in 2003. Because the average number of employees during that time period was 737 and 676 respectively, the overall annual turnover rate had been significantly reduced from 89.6% in 2002 to 56.7% in 2003 (Table 2). In other words, after the intervention in 2002, the company successfully retained almost 40% of the employees it would have lost based on the 2002 trend. The overall turnover rate of this hotel company, especially in 2003, is remarkably low considering that about 75% of the hotel employees are part-time workers who in general have relatively lower job stability and have higher turnover rate than full-time workers.

Table 2 Statistic of Employee Turnover

YEAR	2002		2003	
Average number of Employees	737		676	
	FT employees 178 (24%)	PT employees 559 (76%)	FT employees 181 (27%)	PT employees 495 (73%)
Number of Terminations	660		383	
	FT employees 62	PT employees 598	FT employees 34	PT employees 349
Turnover Rate	89.6%		56.7%	
	FT employees 34.8%	PT employees 107%	FT employees 18.8%	PT employees 70.5%

FT, Full Time; PT, Part Time.

The results shown in Table 2 indicate that the company significantly reduced turnover rates for both full-time and part-time employees. Between 2002 and 2003, part-time employees' turnover rate dropped from 107% to 70.5% whereas the full-time employees' turnover rate dropped from 34.8% to 18.8%. A 34% decrease in part-time employees' turnover rate is a significant improvement in such a short period of time, but a particular attention needs to be paid to a remarkable reduction in full-time employees' turnover rate. Considering much higher costs involved in hiring and training full-time employees and the loss of knowledge, skills, and experiences of those employees, this significant drop (reducing turnover rate almost by 50%) implies a great contribution to the company's bottom line and human resources.

Result 2: Enhancement in Employees' Satisfaction Level

A total of 99 (2002) and 45 (2003) responses to the exit interview surveys were collected for 2-year period and used for the study. The response rates were 19.6% and 15.5%, respectively.

Table 3 shows overall increase in employee satisfaction across the four departments of the hotel. With very few exceptions, the employees who left the company in 2003 rated their average satisfaction levels across the four categories significantly higher than those from the 2002 survey. Particular attention should be paid to the increase of average satisfaction scores between the 2 years—over a 10% improvement for "supervisor" and over a 13% improvement for "overall opinion of the company as an employer" across the four departments.

Table 3 Changes in Employee Satisfaction Level: Exit Interview Survey Responses (2002– 2003)

Departments	Banquets		Food & Beverage		Conference Services		Lodging Operations		Average	
Year	2002	2003	2002	2003	2002	2003	2002	2003	2002	2003
1. Job/Company	3.24	3.11	3.33	3.37	3.60	3.82	3.34	3.82	3.38	3.53
2. Supervisor	3.61	3.64	3.83	3.83	4.09	4.54	2.96	3.98	3.62	4.00
3. Policies & Procedures	3.39	3.15	3.55	3.44	3.33	3.81	3.13	3.62	3.35	3.51
4. Salary & benefits	2.94	2.94	3.09	3.16	3.07	3.43	2.97	3.63	3.02	3.29
Average	3.30	3.21	3.45	3.45	3.52	3.90	3.20	3.76	3.33	3.58
Overall opinion	3.29	3.45	3.34	3.58	3.47	4.38	3.36	3.82	3.37	3.81

Note: A 5-point Likert Scale was used for measurement (5: "very satisfied" to 1: "very dissatisfied")

At department level, the Conference Services and Lodging Operations departments showed very significant increases while the Banquet and Food & Beverage departments fluctuated. The improvement in employee satisfaction in 1 year is a strong indicator that the management training programs from a year earlier may have produced the desired effects.

Discussion

It is critical to notice that during the study period this lodging company did not experience any major variations in other potentially confounding factors that might have contributed to the significant improvements in the outcome measures (i.e., employee turnover and satisfaction level). In other words, there were no significant changes in the hotel's business operations, employee structure, or compensation and benefit programs during the 2-year period. Hence, the development and implementation of quality management training programs can be considered the only significant intervention that could have led to higher employee satisfaction and lower turnover rate. Current and new managers who received the rigorous training were expected to perform more effectively and efficiently particularly in their human resource-related responsibility. After all, the ultimate beneficiaries from quality management training should be their subordinates who are line employees. The study findings are not subject to a concern of a Hawthorne effect since the line-level employees were basically unaware of the newly implemented management training programs. Consequently, the choice of employee turnover and satisfaction as the outcome measures of the actual benefits and effectiveness of the intervention has high validity. This empirical fact supports the value of creating rigorous management training programs and suggests legitimate criteria to which hotel companies can pay attention when they would like to justify the investment in a certain HRM practice.

By identifying the postintervention reduction of employee turnover and a significant improvement in the employee satisfaction level over the 2-year period, the study results demonstrated the importance of a time lag in measuring the effect of major HRM interventions, such as management training programs for this study.

CONCLUSIONS

This study examined a new HRM intervention developed and implemented by a small, nonbranded lodging company that enhanced employees' satisfaction level and reduced turnover rate. The study was designed to verify the effectiveness of the intervention and to measure the effect on those impacted by the training—the line-level employees—rather than on the managers who actually received the training. The authors looked at the direct employee-related factors as opposed to indirect measures such as management's perception or company profit margin. Because the training intervention was designed for a specific goal—improvement of managers' performance in their HR responsibilities—that is the focus of the study and the measurement over time.

The study result shows that the enhanced quality and rigor of the company's training programs for its managerial workforce significantly increased employee satisfaction level and consequently reduced the turnover rate. The reduction in overall employee turnover rates both for part-time and full-time employees were significant despite the post-9/11 economic factors that may have influenced the employees' intention to depart.

To accomplish this, the hiring process and related training programs were completely revised. Individuals being interviewed were given a more realistic job descriptions based on a thorough job analysis and were hired based on their talent as opposed to "warm body syndrome." Prior to the interview training, the assumption was that a manager's skill in interviewing was intuitive and no training was necessary. Providing line managers with the requisite skills to become good interviewers allowed them to make better decisions about who best fit the organization and their particular team. Making the managers more accountable for the hiring process and ultimately for the turnover of employees heightened awareness and contributed to a sense of ownership and thus lower voluntary terminations. One of the tools that assisted the managers participating interviews was an index or matrix (properly vetted by HR experts and legal advisors) they could use to analyze an applicant's characteristics and suitability for a particular job. By better matching between the new hires and the jobs available, it is expected that the managers help prevent the company from losing its new hires due to a mismatch at the early stage of their employment.

This study also demonstrated that there was a significant improvement in employee satisfaction levels postintervention. In looking at the employees' feedback on exit surveys collected in 2002 and again in 2003, it is obvious that the management training had the desired effect. The overall averages were increased over the 1-year time span in all measured areas.

This study has the following strengths. First, the data was collected over a 2-year period in order to avoid the simultaneity problem raised by existing literature. The new and revised management training programs were conducted in 2002 followed by its impact measurement during the period of 2002–2003. Second, the study has strong internal validity because there were no major changes in other confounding factors of the hotel that could have contributed to the changes in the outcome measures between the 2 years. Also, the study was based on data acquired directly from line employees' turnover trends and their responses to exit interview surveys. Contrary to the previous studies that utilized the managerial workforce as study subjects, this study tried to measure the impacts of a management training programs from the perspectives of employees who are the ultimate beneficiaries (or victims) of those practices.

As with most research, there are some limitations that must be noted. First, a convenience sample was used for the study. Because of the nature of the study, one lodging company that initiated key training programs for its managerial personnel was chosen. Second, the response rate from exit surveys is less than 20% for both years due to the characteristics of mail surveys. While still valid, the authors feel obligated to point this out. Third, even though the study design provides relatively high internal validity to the research results, employee turnover still can be affected by external factors, particularly factors over which the company or managers have no or very little control. For example, economic conditions could influence employee turnover. Considering that 2002 was a period of economic decline in the lodging industry due to the terrorist attack on New York City on September 11, 2001, the economy was rebounding by late 2002 and early 2003; these events may have affected the results. Finally, the employee sample of the study is heavily skewed to part-time workers, who are mainly college students. This is not the case with most hotels in major markets. In order to validate the study results, further study is needed to see if the findings can be applied to the industry in general.

Despite the limitations of the study, it does demonstrate the power of HRM practices and that an ROI can be defined for the programs that HR creates. This could be the most positive aspect of the study and have the potential for the greatest future impact. The study results empirically support the significance of effective management training programs in enhancing employee satisfaction and reducing turnover. Since well-trained managers can be more successful in hiring and retaining employees who are motivated and ready to help the organization, it is crucial for the company to develop and implement rigorous training programs, particularly for its managerial personnel.

The significance of this study is that it demonstrates a method that HR professionals can use to establish the fiscal viability of training programs within their operations. The criticality of this study lies in the dwindling resources available to HR managers, in part due to Baby Boomers exiting the workplace, and the high level of voluntary turnover that seems to be endemic in the lodging and restaurant sectors of the hospitality industry. This study suggests a model for HR practitioners to follow that will enable them to measure and demonstrate the ROI of their management training programs. If training mangers about the hiring process, interview skills, and 21st century workplace issues can create a drop in employee turnover as seen in this study, it should be replicated in hospitality properties of all sizes and types. The demonstrated results are significant and can be used to build efficiencies for all hospitality operations.

REFERENCES

1. Allen, D., Shore, L. and Griffeth, R. 2003. The role of perceived organizational support and supportive human resource practices in the turnover process. *Journal of Management*, 29(1): 99–118.

2. Batt, R. and Valcour, P. M. 2003. Human resources practices as predictors of work-family outcomes and employee turnover.*Industrial Relations*, 42(2): 189–220.

3. Baumann, M. 2003. Training, rewards help convince workers to stay. *Hotel and Motel Management*, 218(21): 20

4. Carvin, B. 2002. "Retention management and metrics". Retrieved December 11, 2007, fromhttp://nobscot.com/about/exit_interview2.cfm

5. Chevalier, A., Siebert, W. S. and Viitanen, T. 2003. "Wage policy, employee turnover and productivity". Retrieved October 10, 2005, from http://www.ucd.ie/economic/staff/achevalier/web/turnove r4.pdf

6. Cho, S., Woods, R. H., Jang, S. and Erdem, M. 2006. Measuring the impact of human resource management practices on hospitality firms' performances. *Hospitality Management*, 25(2): 262–277.

7. Development Dimensions International, Inc. 2002. "Reducing turnover". Retrieved June 22, 2006, from:http://www.ddiworld.com/pdf/ddi_reducingturnover_rr.pdf

8. Gomez-Mejia, L. R., Balkin, D. and Cardy, R. L. 2007. *Managing human resources* , 5th ed., Upper Saddle River, NJ: Pearson Prentice Hall.

9. Harris, M., Tang, K. and Tseng, Y. 2002. "Optimal employee turnover rate: Theory and evidence". (Working Paper No. 19/02, Melbourne Institute of Applied Economics and Social Research)

10. Hotel News Resource. 2000. "New Castle Hotel cuts employee turnover to just 57 percent". Retrieved December 5, 2007, from http://www.hotelnewsresource.com/article3860.html

11. Solutions, H R. 2003. "Exit interviews". Retrieved October 7, 2005, fromhttp://www.hrsolutionsinc.com/ov/exit_interviews.htm

12. Humber, T. 2003. "Lowering turnover, increasing morale: The search for the right attitude". Retrieved May 10, 2008, fromhttp://hr.monster.ca/hrreporter/lowering_turnover

13. Huselid, M. 1995. The impact of human resource management practices on turnover, productivity, and corporate financial performance. *Academy of Management Journal*, 38(3): 635–672.

14. Iverson, R. and Deery, M. 1997. Turnover culture in the hospitality industry. *Human Resource Management Journal*, 7(4): 71–82.

15. Kalleberg, A. L. and Moody, J. W. 1994. Human resource management and organizational performance. *American Behavioral Scientist*, 37(7): 948–962.

16. Lee, S., Heard, A. and Koh, W. 2000. "The impact of human resource management practices on employee turnover". Retrieved October 7, 2005, from http://www.fba.nus.edu.sg/fba/mscphd/Index0005.htm

17. Lloyd, R. 2000. "Perceptions of employee turnover". Retrieved October 5, 2005, fromhttp://www.destinationkm.com/articles/default.asp?ArticleID=577

18. Love, G. 2004. Mine employee data to reduce turnover. *Canadian HR Reporter*, 17(7): 13

19. Personnel Policy Service. 2003. "Exit interviews: Constructive feedback, smoother transitions". Retrieved September 30, 2005, from http://www.ppspublishers.com/articles/exit_interviews.htm

20. Pinkovitz, W., Moskal, J. and Green, G. 2002. "How much does your employee turnover cost?". Retrieved October 5, 2005, from http://www.uwex.edu/ces/cced/publicat/turn.html

21. Poe, A. 2003. Keeping hotel workers: It takes more than money to retain lower-paid employees. *HR Magazine*, 48(2): 91–93.

22. Simons, T. and Hinkin, T. 2001. The effect of employee turnover on hotel profits: A test across multiple hotels. *Cornell Hotel and Restaurant Administration Quarterly*, 42(4): 65–69.

23. Smith, H. and Watkins, W. 1978. Managing manpower turnover costs. *Personnel Administrator*, 23(4): 46–50.

24. Sturman, M. and Trevor, C. 2001. The implications of linking the dynamic performance and turnover literatures. *Journal of Applied Psychology*, 86: 684–696.

25. Sweeney, T. 2002. Exit interviews. *Credit Union Management*, 25(11): 38–39.

26. Tracey, J. B. and Hinkin, T. R. 2006. The costs of employee turnover: When the devil is in the details. *Cornell Hospitality Report*, 6(15): 4–13.

27. Tracey, J. B. and Hinkin, T. R. 2008. Contextual factors and cost profiles associated with employee turnover. *Cornell Hospitality Quarterly*, 49(1): 12–27.

28. Woods, R. and Macaulay, J. 1987. Exit interviews: How to turn a file filler into a management tool. *Cornell Hotel and Restaurant Administration Quarterly*, 28(3): 38–46.

Chapter 10

How Individual Beliefs Impact Individual Performance

Carsten Schäfer

Department of Business Administration and Management, University of Economics, Prague, 130 67,Czech Republic

ABSTRACT

This research-paper examines how the beliefs of individuals influence their individual performance. In this context beliefs are subjective doctrines, or rather subjective opinions held by one or more individuals. These doctrines and opinions areclassified as either positive or negative. To measure the deviation of performance of individuals with differentbeliefs, a classroom experiment was conducted. The major finding was a highly significantly improved qualitative performance of 44 %, among individuals with positive beliefs. The findings of the paper imply several approaches for organizational management and psychological/behavioural economics.

Keywords: Performance, Beliefs, Field-Study, Influence, Suggestion, Behavioural Economics, Economic Psychology

1. INTRODUCTION

It's all about beliefs. What does this statement mean? What are beliefs? What are the advantages or disadvantages of beliefs? These questions will all be answered in this paper. These situations are well known: we are faced with an important task that needs to be completed. It might simply be a shortlasting task such as an employment assessmentor an entrance examination. Or it could just as well be the solution to a problem within one's (own) company that is essential for personal progress. A convincing performance has to be delivered in front ofa smaller or larger crowd. Instructors know this problem just as well as musicians, actors, freelancers, employees, professional athletes or managers do. The task may be part of a much greater sphere as well, such as a medium-term or long-term goal in life. Still, who or what determines individual performance in respective tasks or for personal goals or goals set by others? Why do results differ between people with supposedly the same skills? Why do some people achieve their goals effortlessly, while others have to struggle to reach them or even never achieve their goals? The answer to all these questions lies mostly within oneself. This saying is one of these beliefs. Beliefs are an individual's doctrines, subjective opinions and solidly anchored convictions about her- or himself. These doctrinesand subjective opinions either negatively or positively influence actions, progress and performance. This idea applies to companies in regards to their quality management, just as much as to every individual personally. In contemporary economics and psychology, the subject of beliefs and performance has notbeen studied deeply yet.[1]discussed motivation and evaluations as ways to improveperformance in their article, "Effects of goal difficulty, goal setting-method and expected external evaluation on intrinsic motivation." The subject of extrinsic and intrinsic monetary motivation was examined by[2]and[3] in their articles "Extrinsic and intrinsic motivation" and "Pay enough or don't pay at all."[4] discussed selective perceptions of individually preferred information, or rather, preferential reception of information that supports individual preferences in connection with cognitive dissonance. All these factors are largely exogenous. This article on the other hand will discuss endogenous factors (endogenous regarding the individual). These endogenous factors, known as beliefs, have not been extensively researched to date. The individual should not be affected by exogenous influences such as incentives or selective perception, but rather by their inner self. To achieve this, it is nevertheless necessary to initially suggest beliefs exogenously before they become part of the individual's credo and therefore, an endogenous factor. The experiment explained below fills this gap. The extent to which subjective beliefs influence the performance of each individual will be researched. Each person is responsible for his or her own success, be it a top manager, a staff member or the company as a whole. It will become clear which factors shouldto be focused on more in order to activate the existing potential.

2. THEORETICAL PRINCIPLES

2.1. Terminology

The following terms used in this articles and their definitions will be elaboratedon at the end of the text: beliefs in the framework of this article are subjective doctrines, or rather subjective opinions held by one or more individuals. Evaluation is the appraisal or evaluation of a performance. Evaluation can appraise a performance either ordinally or nominally. According to[5], confidence means the trust an individual has in her or hisown performance. Overconfidence describes an excessive, unjustified trust that an individual shows in her or his own performance. The individual overrates her- /himself. Low-powered incentives, or rather low-powered rewards are (in this experiment) rewards in the form of evaluation or feedback. The evaluation or feedback should be positive if possible. A low-powered reward differentiates itself from a monetary reward mostly by itscharacter.[2]discussed the term "low-powered reward."[6]in their article *Large Stakes and Big Mistakes*introduced the term "choking under pressure" regarding rewards. This describes the situation where an individual starts to think a lot about the task. The reward entices them to complete the task particularly well but as a result, they cannot concentrate optimally during the time period in which the given task is to be completed.

2.2. Current Research Results

Recent research results in economics and psychology willbe examined below. The text will focus on findings in the fields of motivation, evaluation and confidence, or overconfidence (self-assessments or self-over assessments). As previously mentioned, no findings in the field of beliefs are known to the author to have been published.

[1]discussed evaluation, the degree of difficulty of a task and intrinsic motivation in their article, "Effects of goal difficulty, goal setting-method and expected external evaluation on intrinsic motivation" . They assumed that there is a connection between evaluation and intrinsic motivation, or rather between the degree of difficulty of a task (in this article, it would correspond with positive or negative beliefs) and intrinsic motivation. An experiment was conducted where the subjects were required to assemble helicopter models. No relationship between evaluation and intrinsic motivation or degree of difficulty of a task and intrinsic motivation was found. Nevertheless,[1] determined that the performance was very high for difficult tasks, or for tasks which are difficult to complete.

[2], in their article"Intrinsic and Extrinsic Motivation," considered evaluation to bea low-powered reward or low-powered incentive (see terminology). An individual is basically interested in (if possible, positive) information about her-/himself. Evaluation as a reward therefore differs significantly from a monetary reward, which is not considered to be a low-powered reward.[2]drew upon the article "Intrinsic Motivation" (1975) by Deci. Unlike[1],[2] found that intrinsic motivation is positively influenced by evaluation or feedback. They justify this with the fact that an individual is more interested in receiving positive than negative information about

her-/himself. In the context of rewards,[2] also found that under certain conditions, an individual may find a task with a high reward less attractive. The individual connects the high reward with a difficult task, which implies more effort for the individual. It can be deduced, therefore, that individuals generally avoid strenuous efforts. The authors postulate that rewards can provide a short-term incentive, but diminish motivation for the future. In their article,[2] also address the individual abilities of the individual. When an individual feels unsure about her/his own abilities, intrinsic motivation sinks with the increasing degree of the reward. This means the higher the reward is, the lower the individual's intrinsic motivation is in this situation. Furthermore[1]agree with[7] when they found that a reward couldinspire an individual to complete a task solely for the reward. The task consequently tends to be exploited and lost its original meaning.[6]introduced the term "choking under pressure" regarding rewards in their article "Large Stakes and Big Mistakes," which describes how a task can prompt an individual to think too much about thattask. The reward prods the individual to complete the task particularly well, but then, (s)hecan no longer concentrate optimally during the time period in which the task is to be completed. The individual begins to ask, "What will happen if I get a bad evaluation? What will they think of me?" As previously described, the individual wants to avoid negative information about her-/himself, or in this case, receive the large monetary award. This phenomenon of the exploitation of an assignment is what the authors call "choking under pressure."[1]found that performancesof difficult tasks dropped significantly.[7]as well as[2] described how rewards that are too high can lead an individual to complete the task solely for the reward. The task is hence also exploited. In connection to the size of the reward,[3] nevertheless found in their article "Pay enough or don't pay at all" that a too low monetary reward lowers the performance also. It therefore seems to be quite important that the optimal size of the reward is carefully considered. The reward should neither exploit the task ([6],[7],[2]), nor should it lower performance. A non-monetary reward in the form of feedback or evaluation (low-powered reward) seems to be appropriate in this situation ([2]).[5]looked at individuals' self-assessments in their article "The Trouble with Overconfidence." They found that on average, individuals rated their achievements too high for easier tasks. Furthermore,[5] found that on average, individuals rated their achievements too low for more difficult tasks. The tasks definedby[5] as simpleor difficult correlate with positive or negative beliefsin this paper.

2.3. Objectives of this Study

As already noted in Section 2.1, the effects of subjective beliefs on performance have not yet been studied in any detail, neither in the fields of (social) psychology nor in economics. The goal of this study is to provide a point of entry for research of in to thesecircumstances and to reveal possible effects. The theoretical principles described abovewill be used to compile variables and formulate hypotheses and secondary hypotheses. The main hypothesis states:

Subjective beliefs influence individual's performance.

This means that positive subjective beliefs of an individual can positively influence their individual performance. For this purpose, a lecture hall experimentwas conducted. The test subjects were split into four groups and each person was given a task to be completed on paper.

Additionally, this article will respond to secondary hypotheses regarding the collected variables. These variables are aligned with the publications presented in Section 2.1. The presented secondary hypotheses are as follows:

- *Evaluation and beliefs influence intrinsic motivation.*
- *The average performance in a treatment with evaluation turns out worse in general.*
- *Individuals assess their performance too high (low) compared to the average for easy tasks (difficult tasks).*

The variables examined within the framework of the experiment are gender, overconfidence and intrinsic motivation.

3. EXPERIMENT

3.1. Design of the Experiments

The following experiment will investigate the influence of positive beliefs (the test subject believes that the given task is particularly easy) and negative beliefs (the test subject believes that the given task is particularly difficult) on performance. The task to be completed is to determine the correct amount of ones inseveral tables with different amounts of zeroes and ones. All test subjects were given the identical assignment, although the cover sheets for the work sheets were prepared in four different versions. The test subjects were each given five minutes time to complete the task. A maximum of twenty tables could be completed.

The test subjects were divided into four groups (see table 2 also):

Group 1: Positive beliefs, evaluation: the test subjects were given a worksheet that convinced them that the given tasks were easy and that there would be a random evaluation or feedback at the end.

Group 2: Positive beliefs, no evaluation: the test subjects were given a worksheet that convinced them that the given tasks were easy and that there would be no random evaluation or feedback at the end. No information regarding a possible evaluationwas provided.

Group 3: Negative beliefs, evaluation: the test subjects were given a worksheet that convinced them that the given tasks were difficult and that there would be a random evaluation or feedback at the end.

Group 4: Negative beliefs, no evaluation: the test subjects were given a worksheet that convinced them that the given tasks were difficult and that there would be no random evaluation or feedback at the end. No information regarding a possible evaluationwas provided.

The following table is an example of the tables that were used within the framework of the experiment:

Table 1. Example of table used in experiment[8]

1	0	0	0	1	1	0	0	1	1	1	1	0	0	1
0	0	0	1	1	0	1	0	1	0	1	0	0	0	1
0	0	0	0	1	1	1	0	1	0	1	0	1	0	0
1	1	1	1	0	0	0	0	0	1	1	1	0	0	1
0	1	0	1	0	1	1	1	1	1	0	0	0	1	1
1	0	1	0	1	0	1	0	1	1	1	1	1	0	0
0	1	0	0	0	0	1	1	1	0	1	1	0	1	0
1	1	1	0	0	0	1	1	0	0	1	1	0	1	0
0	1	0	1	0	1	0	1	0	1	0	1	0	1	0
1	0	1	0	1	0	1	0	1	0	1	0	1	0	1
1	0	1	0	1	1	1	1	0	0	0	1	1	0	1
Amount of Ones:														

The arrangement for the experiment is shown in the following table:

Table 2. Arrangement for the Experiment

	Evaluation	**No Evaluation**
Positive Beliefs	Group 1	Group 2
Negative Beliefs	Group 3	Group 4

The influence that the results from the literature described in Section 2.1 had on the design of the experiment will be described in the following text.

Evaluation was chosen as an incentive within the framework of the experiment, since it can be considered a "low-powered-incentive" ([9],[2]).

The evaluation served as an incentive to participate in the experiment. This is not in conflict with the results of[1], who found in their article "Effects of goal difficulty, goal setting-method and expected external evaluation on intrinsic motivation" that evaluation does not have any effect on intrinsic motivation. In the present study, however, evaluation was intended to serve only as an incentive. This incentive pre-exists as the individual is interested in receiving (hopefully positive) information about her-/himself([9],[2]). Evaluation is therefore to be understood as the provider of extrinsic motivation. The time component is irrelevant in the following experiment, as the given task may only be completed once. The finding that the reward can provide a short-term incentive, but diminish motivation in the future ([2]) is irrelevant here. The test subjects for this experiment were given information about their individual abilities before the beginning of the experiment. They received the information as positive or negative beliefs. One half of the test subjects therefore believed that the task was very difficult, while the other half believe that the given task wasvery simple. This gave the test subjects information about their individual abilities regarding the tasks. The negative correlation between intrinsic motivation and the size of the reward for existing insecurity of the individuals regarding their own abilities described by[2] should have had no effect in this case.[2]as well as[7] also found that a reward can induce an individual to complete a task solely for the reward. To ensure that the test subjects stayed focused on the task during the experiment, no monetary or other kinds of rewards were offered. The "low-powered reward" as described by[2] was used instead.[6]introduced the term "choking under pressure." "Choking under pressure" occurs when an individual thinks too much about the task to be completed. The individual wishes to complete the task successfully because of the reward. In this experiment, the test subjects in the treatments with evaluation were able to ask the following questions: "What will happen if I am evaluated badly? What will people think of me?". The individual wants to avoid negative information about her- or himself. This may cause the "choking under pressure" as described by[6]. If "choking under pressure" does occur, it had to be accepted, as there was no way to avoid it in this experiment. According to[3], performance is diminished if the monetary reward is too low. For this reason, a non-monetary reward was chosen for this experiment.

3.2. Summary

Evaluation in this experiment (in groups one and three) was intended to strengthen the commitmentin the experiment. The evaluations or feedback were implemented as a "low-powered reward" ([9],[2]). The "low-powered reward" kept the individuals focused on the task. The task was thus not exploited ([6]). Evaluation and feedback were used as an external incentive.

In order to limit choking under pressure, only evaluation was as an incentive in groups one and three. In groups two and four, however, no evaluation was communicated to prevent possible choking under pressure. In this manner, the influence of beliefs on the performance in the described arrangement can be optimally determined.

3.3. Compiled Variables

Within the framework of the experiment, the following variables were compiled based on the results described in Section 2.1:

1. Gender (exogenous variable)
2. Overconfidence (endogenous variable)
3. Intrinsic motivation (endogenous variable)

3.4. Task within the framework of the experiment

The tasks shown in this experimentrequire a combination of diligence, quantitative as well as qualitative abilities and intuition. However, diligence as well as quantitative and qualitative abilities areprimary, as the influence of subjective beliefs on performance is intended to be as undistorted as possible. The results are not intended to be influenced by different individual mathematical and or other abilities. Nevertheless, a pure diligence task in connection with suggested beliefs would not becompletely valid. The task most suited for approaching the above mentioned delineations seemed to be the one described in Section 3.1. The task used within the framework of the experiment consisted of determining the number of ones from 20 tables with differentamounts of zeroes and ones.

4. HYPOTHESES

4.1. Hypothesis A

Hypothesis A is the main hypothesis and was the actual goal of the experiment. Because of results described in literature and the supposition of the influence of qualitative beliefs on performance, the following results of the experiment regarding performance are assumed:

Performance group 1 >performance group 2

Performance group 3 >performance group 4

This result tends to refute the explanations of[1] and implies that negative beliefs (the test subject believes that the given task is especially difficult) have a negative influence on performance. Accordingly, positive beliefs (the test subject believes that the given task is especially easy) should have a positive influence on the individual and the collective performance.

The relationship regarding performance between groups 3 and 4 is unclear. Clarificationconcerning how much the evaluation influenced the performance within the given arrangement, or rather, how much choking under pressure ([6]) occurred is necessary in this case. This question will be considered in hypothesis C. Hypothesis A is:

Subjective beliefs influence individual's performance. The performance ofgroup 1 is higher than the performance ofgroup 2. In addition, the performance of group 3 is higher than the performance of group 4.

4.2. Hypothesis B

Hypothesis B is a secondary hypothesis based on the compiled variables.

According to[1], evaluation and the degree of difficulty of a given task (in this experiment, positive or negative beliefs) have no influence on intrinsic motivation.

Within the framework of this experiment, it needs to be clarified whether evaluation and beliefs have an influence on intrinsic motivation. For the time being, it was assumed that this is the case within the framework of hypothesis B. hypothesis B is as follows:

Evaluation and beliefs influence intrinsic motivation.

4.3. Hypothesis C

Hypothesis C is also a secondary hypothesis based on the compiled variables. Evaluation was an external incentive in the form of low-powered rewards ([9],[2]). There is the possibility, however, that they nevertheless elicited the above-cited choking under pressure ([6]). Hypothesis C assumes that the average performance in the treatment with evaluation is generally poorer. Hypothesis C states:

The average performance in the treatment with evaluation is generally poorer.

4.4. Hypothesis D

[5]postulated that for easy tasks (here positive beliefs), individuals overrate their abilities in comparison to their average performance. At the same time, for difficult tasks, individuals underrate their abilities in comparison to their average abilities. Hypothesis D is based on this finding. Hypothesis

D will be verified based on the compiled variable overconfidence. Hypothesis D states:

Individuals assess their abilities too high (low) for easy tasks (difficult tasks) in comparison to their average abilities.

5. ANALYSIS

5.1. Data record and empirical analysis

206 people participated in the experiment. The group distribution within the room during the experiment is shown in the following image.

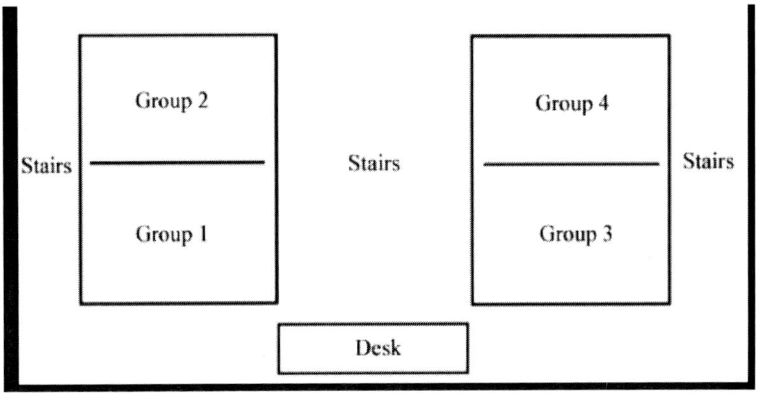

Figure 1. Group distribution within the room during the experiment

The test subject distribution among groups one through four was as follows:

Group 1: 42 participants

Group 2: 49 participants

Group 3: 57 participants

Group 4: 58 participants

45% percent of the participants were male and 55% were female.

The following analysis refers to the tables that were completed correctly by every test subject, henceforth to be called performance. This variable appears to be the most meaningful one, as the number of completed, but not necessarily correctly completed, tables by every test subject is more susceptible to distortion. A test subject could, for example, have randomly entered their results and in that way, completed twenty tables. Analyses were nevertheless carried out indifferent places for the "completed, but not necessarily correctly completed tables" variable as well, henceforth called quantity. This allowed a differentiation between quality, or rather, performance, and quantity. Analyses of the quotients of performance and quantity showed no statistically useful results.

The following results were shown on average (average performance):

Table 3. Average Performance, sorted by groups and beliefs

Group	Mean average
Group 1: positive beliefs /evaluation	6.00
Group 2: positive beliefs / no evaluation	10.96
Group 3: negative beliefs /evaluation	6.21
Group 4: negative beliefs /no evaluation	6.10
Total positive beliefs	8.67
Total negative beliefs	6.15
Mean Average rounded to the second decimal place	

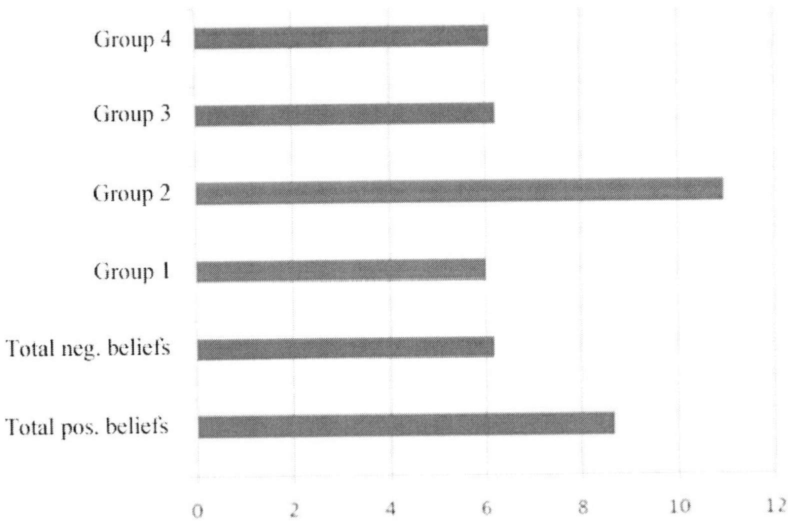

Figure 2. Distribution of performance, sorted by groups and beliefs

It is clear that in the positive beliefs treatment yielded a better average performance than the negative beliefs treatment. The difference is 2.46 correctly completed tables. This indicates an improved performance of 29.07% compared to the treatment of negative beliefs. The difference is even greater when the evaluation criterion during the assessment is considered further. No notable difference was recognized between groups one, three and four. Group two was not evaluated and held positive beliefs. This group achieved an average performance of 10.96 correctly completed tables. Compared to group four, which was also not evaluated but held negative beliefs, the performance improved by 44.34%. Choking under pressure ([6]) is strongly suspected to have occurred in connection with the evaluation in the conducted experiment. The dreaded evaluation may have exploited the task and hence, negatively influenced the performance. Positive beliefs resulted in a performance improvement of 44.34%.

The average performance of all groups was 7.27 correctly completed tables. The median was seven correctly completed tables. There were a minimum of 0 and a maximum of 19 correctly completed tables. The standard deviation was 4.49 (Table 4).

Table 4. Average Performance of All Test Subjects

Statistical size	Value
Average value	7.27
Median	7
Standard deviation	4.29
Maximum	19
Minimum	0
Number of random samplings	206
Values rounded to the second decimal place	

The additional analyses regarding the performance of groups one through four showed the following results:

Table 5. Average performance group one

Statistical size	Value
Average	6.00
Median	5.50
Standard deviation	3.31
Maximum	12
Minimum	0
Number of random samplings	42
Values rounded to the second decimal place	

Table 6. Average performance group two

Statistical size	Value
Average	10.96
Median	11.00
Standard deviation	4.74
Maximum	19
Minimum	2
Number of random samplings	49
Values rounded to the second decimal place	

Table 7. Average performance group three

Statistical size	Value
Average	6.21
Median	6.00
Standard deviation	3.75
Maximum	18
Minimum	1
Number of random samplings	57
Values rounded to the second decimal place	

Table 8. Average performance group four

Statistical size	Value
Average	6.10
Median	6.00
Standard deviation	3.18
Maximum	13
Minimum	0
Number of random samplings	58
Values rounded to the second decimal place	

Descriptive analyses of the average quantity, as described above ("completed tables" as opposed to "correctly completed tables"), are as follows:

Table 9. Average quantity, sorted according to groups and beliefs

Group	Mean average
Group 1: positive beliefs/evaluation	9.71
Group 2: positive beliefs / no evaluation	13.69
Group 3: negative beliefs / evaluation	9.33
Group 4: negative beliefs /no evaluation	8.93
Total positive beliefs	11.86
Total negative beliefs	9.13
Mean average rounded to the second decimal place	

Performance abilities between the treatments of positive beliefs and negative beliefs also differed for the quantity variable. The difference consists of 2.73 completed tables, which translates into a quantitative difference in performance of 23.02%. If the results are differentiated according to the evaluation criterion (treatment of no evaluation), the quantitative difference in performance is 4.76 completed tables. This translates into a performance increaseof 34.77%. Hence, positive beliefs seem to affect the qualitative field more strongly than the quantitative field. The performance improved by 44.34%. Quantity and performance generally seemed to correlate positively ($p < 0.01$ according to Pearson): Test subjects that processed many table also completed many tables correctly.

5.2. Inductive Analysis

5.2.1. Hypothesis A

Median Test

Hypothesis A will first be examined on the basis of a median test. The median test is a nonparametric statistical method for comparing several location parameters. The test will determine whether two or more independent groups distinguish themselves in terms of their central tendencies. The median test describes the probability that two or more independent groups (that do not necessarily have the same composition) come from populations with the same medians. The null hypothesis states that the groups were made up of populations with the same median ([10]).

It was shown that the difference in performance among the specific groups is highly significant (Table 10).

Table 10. Median Test - Hypothesis A – Performance

Group	> Median	<= Median
1	13	29
2	35	14
3	17	40
4	19	39
$\chi2(3)=25.12$, p-value=.00		

The median test shows a significant difference between Groups one through four ($\chi^2(3)=25.12$, p=.00). This means that at least one of the four groups must distinguish itself significantly in its performance from the other groups. Thirteen observations were above the median in groupone, while 29 were under the median. In grouptwo, 35 observations were above the median, and 14 were under it. In groups three and four 17 and 19 observations were below the median, respectively, and 40 and 39 observations above the median, respectively.]

Based on the median tests, it was found that the groups differ significantly in their location parameter. The manner in which the groups differ can be determined using an empirical analysis (Section 5.1). The following figure displays the results of the median test graphically:*Mann-Whitney U Test*

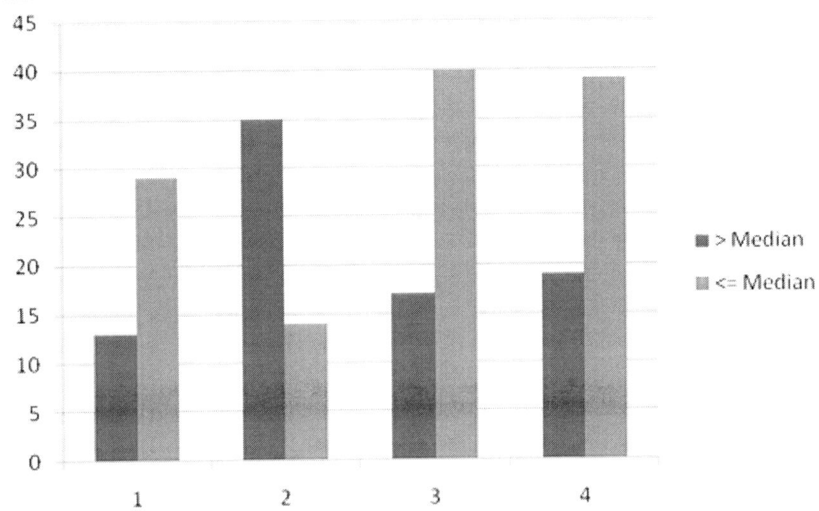

Figure 3. Median test

The Mann-Whitney U Test determines whether two independent random samplings come from the same population ([10]). A comparison between two random samplings can be made regarding the location parameter of a dependent variable (Performance).

Table 11. Mann-Whitney U Test – Hypothesis A – Performance

Belief	Average ranking	Ranking total
positive	120.41	10957.00
negative	90.12	10364.00
Z=-3.63, p-value=.00, positive Beliefs N=91, negative Beliefs N=115		

The Mann-Whitney U Test also shows a significant difference between groups one through four regarding performance (Z=-3.63, p-value=.00).

The Mann-Whitney U Test, broken down by evaluation criteria, also shows that beliefs have a significant influence on performance. It is also evident that performance is influenced by evaluation.

In groups one and four, those with no evaluation, there is a significant difference in performance (Z= -5.07, p-value=.00).

Groups one and three, those with an evaluation, did not show a significant difference in performance Z= -.06, p-value= .95). The results from section 5.1 were therefore confirmed.

Table 12. Mann-Whitney U Test - Hypothesis A – Performance according to Evaluation

Group	Average ranking	Ranking total
1	49.80	2091.00
2	70.49	3454.00
3	50.15	2858.50
4	40.07	2324.00
Group 1, Group 3: Z= -.06, p-value=.95; Group 2, Group 4: Z= -5.07, p-value=.00		

Inductive analyses of the average quantity as described above, ("completed table", not "correctly completed table") are as follows:

Table 13. Mann-Whitney U Test – Hypothesis A – Quantity

Belief	Ranking	Ranking total
positive	123.30	11220.00
negative	87.83	10100.50
Z= -4.265, p-value=.000, positive beliefs N = 91, negative beliefs N = 115		

The Mann-Whitney-U Test regarding quantity also shows a highly significant difference between the positive and negative beliefs treatments (Z= -4.27, p-value=.00). The median test regarding quantity demonstrated a significant difference between groups one through four ($\chi^2(3)$=16.32, p=.00; see Table 14).

Table 14. Median Test - Hypothesis A – Quantity

Group	> Median	<= Median
1	20	22
2	36	13
3	23	34
4	22	36
χ2(3)=16.32, p-value=.00		

ANOVA – UnivariateAnalysis of variance

In order to show the two-factor test set-up (see section 3.1) and to determine more closely the observed reciprocity in the median test and the U test, a two factorialunivariate analysis of variance (ANOVA) was conducted. The ANOVA examines the influence of one or more independent variables on a dependent variable ([11]). Multiple averages for a dependent variable were examined and main effects (beliefs, evaluations), as well as interaction effects (Belief x Evaluation) were considered.

Table 15. ANOVA – Hypothesis A – Performance

Ind. variable	df	Average square	F	p-value
Constant	3	10846.66	758.46	.00
Belief	1	273.13	19.10	.00
Evaluation	1	298.00	20.84	.00
Belief x Eval.	1	324.89	22.72	.00
R2 = .23 (corrected R2 = .22)				

It is clearly recognizable that the averages in groups one and three differ significantly (p-value = .00). The interaction effect of beliefs and evaluation ("Beliefs x Evaluation" variable) is also highly significant (p-value = .00). Also according to ANOVA, at least one average in groups one through four must differ significantly from the other averages. The following imageillustrats this:

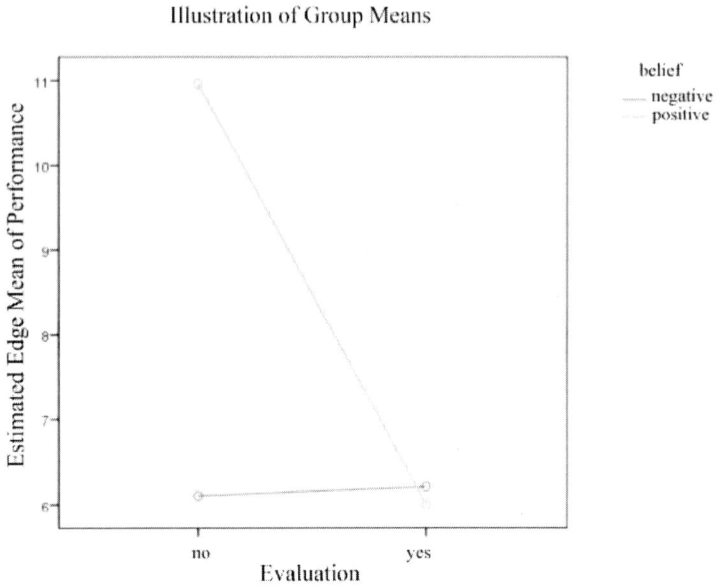

Figure 4. Univariate Analysis of Variance–ANOVA- independent variable: Performance

The imageabove clearly shows that group 2 achieved the highest performance (positive beliefs, no evaluation).

The following is an ANOVA regarding the average quantity, as described above:

Table 16. ANOVA – Hypothesis A – Quantity

Ind. variable	df	Average square	F	p-value
Constant	1	21981.46	1400.99	.00
Belief	1	334.91	21.35	.00
Evaluation	1	161.98	10.32	.00
Belief x Eval	1	243.04	15.49	.00
R2 = .19 (corrected R2 = .18)				

Regarding quantity, it is clearly recognizable that the averages in groups one and three differ significantly as well (p-value = .00). The interaction effect of beliefs and evaluation ("Beliefs x Evaluation" variable) is also highly significant (p-value = .00).

5.2.2. Hypothesis B

Hypothesis B was tested for independent random samplings with a t-test. The t-test tested the averages of two random samplings for significant differences ([11]).

Table 17. T-Test - Hypothesis B – for all groups

Variable	T-statistic	p-value	Difference	SD difference
Motivation	-1.59	.14	-.42	.27
Group evaluation No: n=106, evaluation Yes: 98				

No significant influence of evaluation on intrinsic motivation could be determined. The results of[1] were therefore confirmed in this experiment (T= -1.59, p= .14). Hypothesis B could not be confirmed regarding the evaluation. The arrangement where no evaluation was mentioned had a higher average motivation value than the arrangement where evaluations were mentioned (averages = 3.59 to 3.53). Nevertheless this difference is not significant.

When a t-test sorted by gender of the test subjects is conducted, however, a significant influence of evaluation on motivation is seen (p-value = .01).

Table 18. T-Test – Hypothesis B – sorted by gender

Variable	Gender	T-statistic	p-value	Difference	SD difference
Motivation	male	-2.71	.01	-1.09	.40
Motivation	female	6.31	.59	.215	.34
male: n = 90; female: n = 114, no gender given: n = 2					

Male test subjects seemed to be influenced by evaluation regarding their motivation. This finding in the experiment was not observed among women.

When testing the influence of beliefs on intrinsic motivation, the following results were found:

Table 19. T-Test - Hypothesis B – for all groups

Variable	T-statistic	p-value	Difference	SD difference
Motivation	.20	.84	.05	.27
Belief positive: n= 91, Belief negative: n= 115				

There is no significant influence of beliefs on motivation (p-value = .84). Regarding beliefs hypothesis B could not be confirmedeither.

Calculations that included the gender factor did not reveal significant results (male: p-value = .66, female: p-value = .81).

5.2.3. Hypothesis C

As already described in section 4.2.1, evaluation influenced average performance rather negatively. To examine this fact further, a linear regression analysis that includes all other collected independent variables was conducted here.

Unlike the correlation calculation that simply determines the strength of the relationship between two variables, regression analyses serve to reveal the type of relationship. In the process, the value of a dependent variable is then to be predicted precisely, based on several independent variables ([11]).

The linear regression equation for this experiment is as follows:

$$Performance = \beta_0 + \beta_1 Motivation + \beta_2 Gender + \beta_3 Evaluation + \beta_4 Belief + \varepsilon \quad ([12])$$

The achieved performance in turn forms the dependent variable. Intrinsic motivation, gender, evaluation and belief incorporate the independent variables. Beta 0 through beta 4 represent the influence of the respective independent variables on the dependent variable. Beta ranges between zero and one. Epsilon represents the existing error term.

The result of the linear regression analysis was as follows:

Table 20. Linear Regression Analysis– Hypothesis C – Performance

Variable	Beta	Standard error	Beta stand.	T-statistic	p-value
Constant	8.02	.79		10.13	.00
Motivation	-.14	.15	-.06	-.90	.37
Gender	-.44	.57	-.06	-.84	.40
Evaluation	-2.20	.56	-.21	-3.89	.00
Belief	2.37	.56	.28	4.22	.00
F=9.04, p=.00, R2 corrected=.14					

As shown in the table above, the variables evaluation and belief have a highly significant influence on the dependent variableperformance (p-value= .00). This finding corresponds with the previously found results and effects. Intrinsic motivation (p-value = .37) and the gender of the test subject (p-value = .40) do not influence the performance strongly or significantly.

A strongly significant influence of (positive) beliefs on the achieved performance can be identified. According to the regression analysis conducted here, test subjects with positive beliefs correctly complete 2.37tables moreon average than test subjects with negative beliefs (β = 2.37). In hypothesis A a highly significant influence of beliefs on performance wasalready found.

The findings of this experiment regarding evaluation were confirmed here. Test subjects who were to be evaluated correctly completed 2.20 tables less on average (β = 2.20, once again according to the regression analysis conducted here). Hypothesis C is therefore confirmed.

Nevertheless, the here underlying moderate model adjustment ($R^2 = .15$, R^2 corrected=.14) must be considered when interpreting these results.

Figure 5 graphically shows the regression model and its errors of estimation.

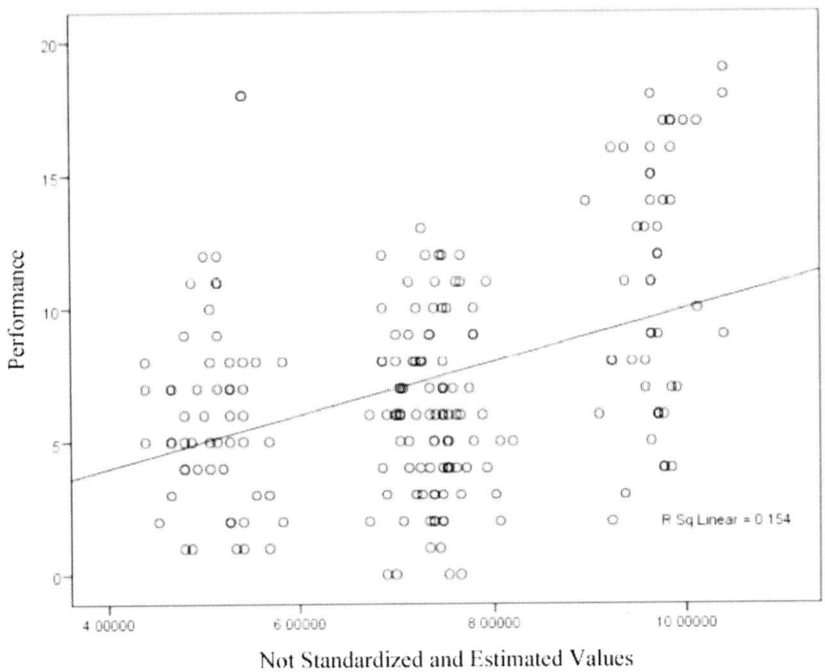

Figure 5. Scatter Diagram of the Linear Regression Analysis–Performance

The following is a linear regression analysis using the dependent variable quantity, as described above ("completed tables", not "correctly completed tables"):

Table 21. Linear Regression Analysis – Hypothesis C – Quantity

Variable	Beta	Standard error	Beta stand.	T-statistic	p-value
Constant	10.02	.81		12.36	.00
Motivation	-.10	.15	-.05	-.67	.50
Gender	-.72	.59	.08	1.22	.22
Evaluation	-1.50	.58	-0.17	-2.59	.01
Belief	2.57	.58	.29	4.47	.00
F=7.93, p=.00, R2 corrected=.12					

As shown in Table 21, not only was the performance influenced significantly, but the quantity of the variables evaluation (p-value = .01) and belief (p-value = .00) and the constant variables of the models, were as well. The intrinsic motivation (p-value = .50) and the gender of the test subjects (p-value = .22) did not significantly influence the performance.

Also in regards to quantity, a highlysignificant influence of beliefs on the dependent variable"completed tables" was noted. Test subjects with positive beliefs completed 2.57 more tables on average in this regression analysis than the test subjects with negative beliefs ($\beta = 2.57$).

The results regarding the evaluation concur with the results of the dependent variable performance: evaluated test subjects completed 1.50 fewer tables (β=1.50)on average.

However, regarding the quantity, the underlying moderate model adjustment R^2must be considered ($R^2 = .14$, R^2 corrected= .12).

5.2.4. Hypothesis D

Hypothesis D was tested using the following cross-classified table:

Table 22. Cross-Classified Table – Overconfidence x Performance Categories

Belief	Overconfidence	Best range 25%	Better range 50%	Worse range 50%	Worst range 25%	Total
Negative	1	5	4	3	2	14
	2	22	18	12	11	63
	3	8	4	7	8	27
	4	0	1	1	3	5
	Total	35	27	23	24	109
Positive	1	3	4	0	4	11
	2	17	9	18	7	51
	3	6	8	5	8	27
	4	0	0	0	0	0
	Total	26	21	23	19	89
Positive Belief: Pearson's Chi-Square = 11.29, p=.08 Negative Beliefs: Pearson's Chi-Square = 8.83, p=.45						

The above-presented cross-classified table is to be interpreted in the following manner:

In the field of positive beliefs, eleven test subjects ranked their abilities to be within the top25%. In fact the abilities of three of the eleven test subjects were assessed correctly. 51 test subjects ranked their abilities to be in the better 50%. Only 17 of these 51 people actually performed in the way they had ranked themselves. In the field of positive beliefs, none of the test subjects ranked themselves to be in the worst 25%. This distribution leads to the premonition that the "overconfidence"[5] phenomenon did in fact arise here. Individuals with simple tasks or positive beliefs seemed to rank their abilities too wellin relation to the average. However, the chi-squared test showed a level of significance of only .08 (p-value= .08) in the field of positive beliefs. The hypothesis is therefore significant to a level of p= .10 only.

The field of negative beliefs shows a different picture. Estimations of the performance levels compared to the average performance are generally worse than those in the area of positive beliefs. Five test subjects rated their abilities in the worst area. 14 ranked themselves in the upper half of all test subjects. The chi-squared test confirmed this finding. A level of significance of only .45 (p-value = .45) was reached. The finding is thus not significant.

Hypothesis D stated that individuals ranked their own abilities in easy tasks or with positive beliefs too well compared to the average. Individuals rank their own abilities in difficult tasks or with negative beliefs too low compared to the average ([5]). This hypothesis was only partially confirmed in the experiment performed here.

6. CONCLUSIONS

This study examined to what extent subjective opinions, known as beliefs, significantly influence individual´s performance. The paper starts with the thoughtthat beliefs have the power to influence everyone. This thoughtis in turn an opinion, a belief. The intent was to look at this opinion more closely. If this opinion could be confirmed, it would be important for all circumstances. In economics, these results could be particularly useful for business coaching, but also for positively influencing the abilities of all individuals within an organization. For this reason, an experiment with a total of 206 participants was conducted. The task was to determine the correct amountof ones in 20 tables consisting of differing amounts of ones and zeroes. The performance was rated based on the number of correctly completed tables. The quantity, meaning the number of completed tables (not "correctly completed tables"), was also considered. Preliminary review of the data had already determined that the performance of people in the group with positive beliefs differed greatly from the performance of the people in the group with negative beliefs. Several statistical tests (t-Test, Mann-Whitney U Test, Median Test, ANOVA, Regression Analysis) revealed that these differences in performance are highly significant (p-value =.00). A difference in abilities regarding performance of approximately 45% was determined. The difference in performance regarding quantity ("completed tables", not "correctly completed tables", see above) was approximately 35%. Positive beliefs therefore had a stronger influence on performance than on quantity. However, these differences arose in the setting of "No Evaluation" and in the "Evaluation" setting, the difference was less clear. Beliefs therefore influence individual abilities very strongly. This influence is highly significant. This finding conflicts with the findings of[1], who found that when the test subjects were given difficult tasks, the performance was especially high. The fact that the time allowed for the shown beliefs to affect the test subjects was very short mustto be considered here. Only one single sentence was given: "This task is very difficult to complete" or "This task is very easy to complete". What kind of results would have occurred if therehad been even more time to consider this thought? What increase in ability could be observed over a longer period of time – a year, ten years, an entire lifetime? What would a person with positive beliefs have achieved by the end of their life, and what would a person with negative beliefs have achieved? An intellectual challenge or the quality of a solution to a task does not depend necessarily and solely on the individual´s professional abilities. It depends much more strongly and thoroughly on the individual´s mental abilities. For now, it is therefore important to beaware of one's own mental abilities and internalize one´s own positive beliefs, as these directly affectthe thoughts and mental models of the individual. Mental models and beliefs furthermore directlyaffect the actions of the individual, as individualsactaccording to their opinions and value beliefs. The more deeply an individual is convinced of a belief, the stronger his/heractionsare influenced by it. This experiment used persuasion to simulate an individual's belief. Long-term, however, it is not persuasion, but rather the endogenous beliefs of the individual that are important. Unlike beliefs, evaluation seemed to influence a performance negatively in this experiment. Not only was the performance influenced negatively here, but the effect of the belief was also undermined. Evaluation seemed not to actas an ability-strengthening element in this experiment. As shown, an individual should not let her-/himselfbe distracted from the task by evaluation, but should rather focus on her/his positive beliefs. In this way, the "choking under pressure" phenomenon ([6]) can be avoided. Choking under pressure occurred in this experiment and made it clear that evaluation (in this experiment) negatively influences performance.

The findings of[6], where an evaluation was found to have no influence on intrinsic motivation, were confirmed in this experiment. The difficulty of the task (positive beliefs versus negative beliefs) did not influence intrinsic motivation in this experiment either. This also corresponds with the results of[1], published in their essay entitled "*Effects of goal difficulty, goal setting-method and expected external evaluation on intrinsic motivation.*" In both cases (group 1 and group 3), the fact that intrinsic motivation was provided through self-assessment by the test subjects themselves mustto be considered. The results are less significant than the results regarding performance (Hypothesis A).

In the context ofthe economy and the positive influence of performance on an organization and on the individual within an organization, the given results imply that it is usefulto integrate the belief factor into coaching sessions. Coaching should not merely feature elements such as evaluation and monetary incentives without consideration of inner persuasions. Coaching sessions should rather consider endogenous factors (beliefs) as well. A coaching session based on beliefs can, in its entirety, be used in an organization. The results described here are valid for top management as well as for all other hierarchy levels within an organization. An understanding of mental models and beliefs can also help to significantly improve performance. It is important that the individual is "programmed for success." That does not mean manipulating the individualunder any circumstances. Rather, the point is to show the individual how to use their beliefs in a purposeful manner. In this experiment, the suggestion of positive or negative beliefs served solely to simulate the doctrines or the beliefs of the individual. An individual's positive beliefs ultimately lead to an effective and efficient (better) prevailing mood within an organization. It is important to show the individual how (s)he can use their positive beliefs effectively.

The following five steps can therefore form the basis for a coaching session:

1. Identify desirable outcomes, define goals.

2. Develop beliefs that support achieving these goals.
3. Act according to the formed beliefs and so, act in a targeted manner.
4. Iterative observation of the beliefs to determine whether they are optimally focused on the desired goals.
5. Observe and if necessary, adjust own beliefs ([13], [14]).

If an individual wants to achieve a certain goal, he or she should create a "mindset," which represents the desired results of her/his individual actions. A belief is nothing more than a set or suggested thought pattern that the individual assumes is reality. Therefore, when the individual assumes that "this task is difficult", the belief leads to the corresponding result. This interpretation forms the behaviour of the individual. Whether new assets or new customers are to be gained, or simply leaving the office at 5p. m., the individual's beliefs influence her/hisbehaviour. The result is that the individual's beliefs form his/herresults. Henry Ford described this in his famous quote, "*If you think you can do a thing or can't do a thing, you're right.*" An individual will not investmuch energy for a goal that (s)hethink is unattainable, which means that the (s)he will not achieve that goal. Simply honing one's skills does not suffice. It is important to train beliefs to support the efforts of many large companies that are based on behaviour-based training and monetary incentives. Positive beliefs, as shown here, can lead to better results. Most dictionaries describe a doctrine as the belief that certain things are true. Therefore, it is important that individual beliefs reflect and support the desired performance. Beliefs and mental models are subtle, but powerful. They are subtle because the individual is usually not aware of their effect. They are powerful because they determine where the individual's attention, and therefore their actions, is focused. Mental models and beliefs are also strict and conservative because they trigger an individual to see what they have always seen: the same needs, opportunities and results. This explains why an individual sees what her/his beliefs allow her/him to see; they are only able to do what their beliefs allow her/him to do. How should an individual comport her-/himself? What activities allow an individual to develop her-/himself? What output should be achieved and what standards determine effectiveness? Since the mental models condition the well-tested way of thinking and course of action of an individual or a collective, an organization must reconsider, review and analyse their beliefs. Which beliefs hinder an organization and which beliefsare the cornerstone of the organization can be determined ([13],[14]).

Beliefs areconspicuous in readings or in comparisons ofbiographies of successful figures as well: they direct the necessary positive beliefs and are prepared to act accordingly. These people identify with their tasks. Their beliefs are after all anchored within their performance. Psychologists call this total concentration and the "absorption in an activity in a relaxed manner" a state of flow ([15]). It is not enough to start off from a binary thought pattern (positive versus negative beliefs), as was shown in the described experiment. Rather, it is much more important to develop an holistic thought pattern or a belief system. Optimally, this belief system is oriented toward individual goals. It is of considerable economic value to continue the research in beliefs. More complex models need to be developed to enablemore detailed research into the consequences of beliefs. On this basis, it would be possible to use beliefs more comprehensively and more practically.

REFERENCES

1. Oldham, Greg R.; Porac, Joseph F.; Shalley, Christina E. (1987). "Effects of goal difficulty, goal setting-method and expected external evaluation on intrinsic motivation." Academy of Management Journal 30 (1987) 3: 553-563.
2. Benabou, Roland; Tirole, Jean (2003) "Intrinsic and Extrinsic Motivation." The Review of Economic Studies 70 (2003) 3: 489-520.
3. Gneezy, U.; Rustichini, A. (2000). "Pay enough or don't pay at all." The Quarterly Journal of Economics 115 (2000) 3: 791-810.
4. Hart, W.;Eagly, A. H.; Lindberg, M.J.; Albarracin, D.; Brechan, I.; Merrill, L. (2009): Feeling Validated Versus Being Correct: A Meta-Analysis of Selective Exposure to Information. Psychological Bulletin 135 (2009) 4: 555-588 www.apa.org.
5. Moore, D.A.; Healy, J.P. (2008): The Trouble with Overconfidence. Psychological Review 115 (2008) 2: 502-517.
6. Ariely, D.; Gneezy, U.; Loewenstein, G.; Mazar, N. (2009). "Large Stakes and Big Mistakes." The Review of Economic Studies (2009) 70: 451-469.
7. Condry, J.; Chambers. J. (1978). "Intrinsic Motivation and the process of learning." In: Lepper, M.; Greene, D.: The hidden costs of reward: New perspectives on the psychology of human motivation. New York: John Wiley: 61-84.
8. Abeler, J.; Falk, A.; Götte, L.; Huffman, D.; (2009): Reference Points and Effort Provision. IZA Discussion Paper No. 3939 (2009) www.ssrn.com.
9. Deci, E.L. (1975). Intrinsic Motivation. New York: Plenum.

10. [10] Siegel, Sidney (1987): Nichtparametrische statistische Methoden. Frankfurt am Main: Eschborn.

11. Bühl, Achim (2008): SPSS 16 – Einführung in die moderne Datenanalyse. München: Pearson Studium.

12. Hassler, U.; Nautz, D; (2004/2005): Statistik I und II – Vorlesungsunterlagen. Johann-Wolfgang-Goethe Universität Frankfurt am Main.

13. Lodie, Bob (2006): It's all about Beliefs http://www.smallbusinessbrief.com/articles/inspiration/003753.html

14. Cerminaro, Anthony (2006): "Whether you think you can or can't, you're right" – Henry Fordhttp://www.allbusiness.com/legal/contracts-agreements/3874550-1.html

15. Csikszentmihalyi, M. (2008): Flow: The Psychology of Optimal Experience. New York: Harper Perennial Modern Classics.

Chapter 11

Explaining Intrapreneurial Behaviors of Employees with Perceived Organizational Climate and Testing the Mediating Role of Organizational Identification: A Research Study among Employees of Turkish Innovative Firms

Seçil Bal Taştana , Cem Güçelb

a Marmara University, Faculty of Business Administration, İstanbul, 34180, Turkey

b Başkent University, Ankara, Turkey

ABSTRACT

This study examines perceived organizational climate and organizational identification as potential antecedents of employees' intrepreneurial behaviors. In particular, the study suggests positive relationships between perceived organizational climate components-structural support and recognition- and intrepreneurial behaviors construct. In addition, employees' organizational identification is suggested to have a mediating role on the relationship between organizational climate and intrepreneurial behaviors. The survey of this study is performed among employees working in high performing and innovative firms operating in White Good Manufacturing, Food and Drink, Telecommunication, and Textile industries in Turkey. The obtained data from the questionnaires are analyzed through the SPSS statistical packaged software. Analyses results revealed that both dimensions of organizational climate (structural support and organizational recognition) significantly and positively related to intrepreneurial behaviors and perceived organizational identification mediate the effects of the organizational climate on the intrepreneurial behaviors construct.

Keywords : Intrapreneurial behavior; Organizational climate; Organizational identification; Innovative organizations

1. INTRODUCTION

The current study aims to explain employees' intrepreneurial behaviors by focusing on the organizational climate dimensions of structural support and organizational recognition and perceived organizational identification. Based on the literature review and conceptualizations of the concepts, it is found meaningful to examine the relationship between organizational climate, organizational identification and intrepreneurial behaviours in high performing innovative organizations located in Turkey. Thus, the main concerns of this study are understanding intrepreneurial behaviours in innovative organizations and examining the relations of perceived organizational climate and organizational identification as potential antecedents.

2. LITERATURE REVIEW AND HYPOTHESES

2.1. Intrapreneurial Behavior

Intrapreneurship concept has been linked to the entrepreneurial orientation of an organization and it has its roots in entrepreneurship literature, even though intrapreneurship as a concept has lately been mentioned in the management literature (Antoncic and Hisrich 2003; Heinonen and Korvela, 2003). In both literatures, intrapreneurship is seen as important for organizional survival, growth, profitability, and renewal (e.g., Zahra 1996; Heinonen and Korvela, 2003; Alpkan, Bulut, Günday, Ulusoy and Kılıç, 2010; Sijde, Veenker and During, 2013).

In broad definition, intrapreneurial behavior is described as entrepreneurship within an existing organization (Antoncic, 2007). The domain of intrapreneurial behavior has been defined in various concepts such as intrapreneuring (Pinchot, 1985), internal corporate entrepreneurship (Jones and Butler, 1992), corporate entrepreneurship (Stopford and Baden-Fuller, 1994; Antoncic and Hisrich, 2004) and corporate venturing (Vesper, 1990). Stevenson and Jarillo (1990) saw intrapreneurship as a process by which employees inside organizations pursue opportunities without regard to the resources they currently control. A long with this definition, Vesper (1990) implied that intrapreneurial behavior was doing new things and departing from the customary to pursue opportunities. Besides, Antoncic and Hisrich (2001) addressed that intrapreneurship was employees' spirit of entrepreneurship within the existing organization. In addition, it is mentioned that intrapreneurial behavior is initiated in established organizations for purposes of profitability, strategic renewal, fostering innovations, gaining knowledge of future revenue streams and international success. It was indicated that the intrapreneurs acted like entrepreneurs in that they realize their own ideas without being the owner of the enterprise (Cunningham and Lischeron, 1991). Thus, intrapreneurial behavior can be defined to mean an entrepreneurial way of action in an existing organization. In their precise conceptual definition of intrapreneurial behavior, Antoncic and Hisrich (2003) defined the concept as the pursuit of creative or new solutions to challenges confronting the firm, including the development or enhancement of old and new products and services, markets, administrative techniques and technologies for performing organizational functions, as well as changes in strategy, organizing, and dealing with competitors (p.8).

On the other side, Heinonen (1999) implied that the basis of intrapreneurial behavior was recognizing an opportunity, exploiting it and trusting that exploiting an opportunity in a new way that deviates from previous practice will succeed and support the realization of the organization's goals (as also cited by Heinonen and Korvela, 2003, p.3). Antoncic and Hisrich (2001) classified intrapreneurial behavior into four dimensions: new business venturing, innovativeness, self-renewal and pro-activeness. Antoncic (2007) indicated that wew business venturing dimension refers to the creation of new businesses related to existing products or markets without regard to the level of autonomy or size; innovativeness dimension refers to product and service innovation with an emphasis on development in technology; self-renewal dimension reflects the transformation of organizations through renewal of the key ideas on which they are built; and proactiveness dimension includes initiative and risk taking, and competitive aggressiveness that are reflected in activities of top management (p.311).

Moreover, Antoncic and Antoncic (2011) addressed that intrapreneurship occurs on two levels: the level of the organization and on the level of the individual. Referring to the view of Antoncic and Antoncic (2011), in the current study we focus on both of the levels and more specifically on the perceived organizational climate (organizational level) and organizational identification (individual level). Supporting and recognizing organizational climate by the organization and organizational identification are suggested to be beneficial for intrapreneurship (Carrier, 1996).

2.2. Antecedents of Intrapreneurial Behavior

Several researchers have attempted to understand the elements that stimulate or effect intrapreneurial behavior. Areas such as external environment (Zahra, 1993), organization (Schollhammer, 1982; Kanter, 1984; Antoncic and Hisrich, 2001; Antoncic and Zorn, 2004), organizational strategies and research and development activities (Antoncic and Hisrich, 2001; Sijde, Veenker and During, 2013), management activities (Antoncic and Hisrich, 2004;

Fitzsimmons, Dougles, Antoncic and Hisrich, 2005), and organizational culture (Heinonen and Korvela, 2003) have been examined as factors affecting intrapreneurial behavior. As mentioned earlier, intrapreneurship is a process, which occurs in interaction with the environment and the organizational setting and the environment plays an important role in influencing intrapreneurship. Basicly, the literature on intrapreneurial behavior has identified two main categories of antecedents: one pertains to the external environment of the firm, the other to its organizational characteristics. The external environment has been viewed as a determinant of entrepreneurial activity at the organizational level (Miller, 1983; Khandwalla, 1987; Covin and Slevin, 1991). It was indicated that the more dynamic, hostile and heterogeneous the environment, more emphasis the company puts on intrapreneurial activities (Zahra 1993; Heinonen and Korvela, 2003). Researchers explained and predicted intrapreneurship and its outcomes with internal variables and external environment variables by building on contingency theory models (Zahra, 1993; Badguerahanian and Abetti, 1995; Antoncic and Hisrich, 2004; Antoncic, 2007). The other category named as organizational characteristics has been viewed as a determinant of intrepreneurial activity at the individual level. Previous research has focused on characteristics of intra-organizational environments that could represent stimulants or antecedents for intrapreneurship development (Schollhammer, 1982; Kanter, 1984; Antoncic and Hisrich, 2001; Antoncic and Zorn, 2004). Organizational characteristics (communication openness, control mechanisms, environmental scanning intensity, organizational and management support, organizational climate, organizational values) compose the group of predictors of intrapreneurship (Antoncic, 2007).

2.3. Organizational Climate and Intrapreneurial Behavior

Organizational climate of an organization refers to the form of the existing conditions and nature of organizational life perceived by the employees. With a pioneering definition, organizational climate is a set of individual, organizational and environmental character features which gives an identity to the organization by separating it from others, perceived by individuals and has an effect on their behaviors (Friedlander and

Greenberg, 1971). Litwin and Stringer (1974) defined organizational climate as the set of measurable properties of the work environment that is either directly or indirectly perceived by the employees who work within the organizational environment that influences and motivates their behavior (p.13). More particularly, the extant literature has confirmed the links between organizational climate and employee performance, productivity, job satisfaction, organizational commitment, organizational justice, motivation, anxiety, intention to leave, and organizational effectiveness (e.g., Dickson, Resick and Hanges, 2006; Bellou and Andronikidis, 2009, Zhang and Liu, 2010; Holloway, 2012; Rahimic, 2013).

In the literature, organizational climate has been studied as an independent or intervening variable between organizational systems and motivational tendencies variables, and as one of the system determinant, the concept has been indicated as influencing motivational tendencies of the individuals and organizational consequences (e.g., Litwin and Stringer, 1974; Özdemir, 2006; Ergülen, 2011). It was revealed that organizational climate created a sense of belonging for employees and the characteristics of the climate which are internalized by individuals are related to a variety of employee behavioral outcomes. After conducting a research study among employees working in Turkish family-owned organizations, Ergülen (2011) has demonstrated that organizational climate was significantly related to individuals' organizational attachment, organizational identification and organizational commitment. Zhang and Liu (2010) has implied that organizational climate influenced organizational and individual variables. When employee's perceived a supportive climate, they exhibited positive behaviors and identification and when they perceived unpleasant working conditions, they tended to commit counterproductive behaviors. In addition, a study performed in Turkey has also revealed that positive organizational climate perceptions related to positive employee behaviors, however negative organizational climate perceptions related to counterproductive behaviors (Kanten and Ülker, 2013). As further, some other researchers argued that organizational climate promoted positive behaviors in organizations such as organizational citizenship behavior, innovative behavior, creative and proactive behaviors (Patterson, West, Shackleton, Dawson, Lawthom, Maitlis, Robinson and Wallace, 2004; Bindl ve Parker, 2011; Farooqui, 2012; Moghimi and Subramaniam, 2013).

On the basis of that rationality, in the current study, we discussed how perceived organizational climate was related to intrepreneurial behavior and organizational identification. We conceptualized organizational climate as a global construct, covering two basic climate aspects: recognition of the organization and structural support. It is suggested that the two climate aspects could provide a good representation of the organizational climate within an innovative organization, and provide information about how recognition and structural support of the organization are related to individuals' organizational identification and intrepreneurial behavior. Thus, the following hypothesis is generated;

H1: A significant and positive relationship is expected between perceived organizational climate and intrapreneurial behavior.

2.4. The Mediating Role of Organizational Identification on the Relationship between Organizational Climate and Intrapreneurial Behavior

Social Identity theory suggested that an individual's sense of self exerts a significant effect on his or her perceptions, attitude and behaviors (Tajfel and Turner, 1985). Being the pioneer of Social Identity Theory, Tajfel (1978) defined social identity as "that part of an individual's self concept which derives from his or her knowledge of his or her membership to social group (or social groups) together with the value and emotional significance attached to that membership" (p.63). Based on the rationality of Social Identity Theory, Mael and Ashforth (1995) have asserted that organizational identification is a specific form of social identification where the individuals define themselves in terms of their membership in a particular organization (p.310). Hatch and Schultz (1997) have viewed organizational identity as consisting of attributes that members feel are central, distinctive, and enduring, or it may refers broadly to what members perceive, feel, and think about their organization (p.357). Moreover, it was indicated that individuals' evaluation of whether the organization's identity is favorable or unfavorable is based on the individuals' subjective assessment of (a) those subjective factors believed to comprise an organization's identity, and (b) the perceived attractiveness of those compositional factors as they are understood by the organizational member (Dukerich, Golden and Shorteli, 2002, p.509).

The extant literature provides a number of studies which confirmed strong link of organizational identity to several individual work outcomes as well as the particular organizational and individual antecedent to identification (e.g., Miller, Allen, Casey and Johnson, 2000; Mael and Ashforth, 2001; Kreiner and Ashforth, 2004; Harris and Cameron, 2005; Riketta, 2005; Cole and Bruch, 2006). Schrodt (2002) examined the relationship between organizational identification and organizational culture and revealed that employee perception of culture significantly influenced identification in retail sales organizations (p.189). Lee (2004) indicated the link between perceived organizational trust and identification. Cole and Bruch (2006) demonstrated that organizational identification has significant relationships with organizational commitment and turnover intention of employees (p.585). Tüzün and Çağlar's (2009) study confirmed that organizational identity was significantly related with trust construct (p.284). Vondey (2010) has revealed that there were significant relationships among servant leadership, organizational citizenship behavior, person-organization fit, and organizational identification (p.3). Moreover, a recent study conducted in Turkish context has showed that employees' perception of organizational identification mediated that relationship between psychological empowerment and voluntary performance behaviors (Taştan, 2012, p.227).

Though there is a body of findings related to the potential antecedents and outcomes of organizational identification construct, it is noticed that there are insufficient research studies positing its links to employees' intrapreneurial behaviors at work and organizational climate. Therefore, in line with the previous findings, the following hypotheses are generated in this study;

H2: A significant relationship is expected between perceived organizational climate and organizational identification.

H3: A significant relationship is expected between perceived organizational identification and inrapreneurial behavior.

Furthermore, in the current study, it is suggested that organizational identification mediates the relationship between perceived organizational climate and intrapreneurial behavior. The theoretical framework for this argument is derived from social identity theory (Ashforth and Mael, 1989), in which the perception of oneness with or belongingness to an organization defines the individual in terms of the organization. In addition, it is argued that such identification with the organization leads the individual to act in ways that are congruent with that identity (Vondey, 2010). The implication of a mediating effect of organizational identification for this study is that the perceived organizational climate would relate to intrapreneurial behavior with the intervention of organizational identification. Hence, it is suggested:

H4: The perceived organizational identification mediates the relationship between perceived organizational climate and intrapreneurial behavior.

3. METHODOLOGY

3.1. Research Aim

In this survey it is aimed to identify the mediating effect of organizational identification on the relationship between organizational climate and intrapreneurial behavior. To test the hypothesized relationships, a field survey using questionnaires was conducted.

3.2. Sample and Data Collection

The survey of this study was conducted on 210 employees working as lower and middle level managers and as nonmanagers of 5 high performing and innovative firms operating in White Good Manufacturing, Food and Drink, Telecommunication, and Textile industries in Turkey. The firms were selected from the list of Turkish Time Journal (2011). The official report of Turkish Time Journal (2011) listed the "The Most Innovative Firms in Turkey". 36 firms were placed in that report however, 5 of them were identified from different industries and were contacted via email or phone and with personal interviews in order to be informed about the research study. About 250 questionnaires were distributed to the sample group and 210 usable questionnaires could be obtained. Data were analyzed through the SPSS statistical packet program and three proposed hypotheses were tested through regression analyses.

3.3. Instrument

To measure "organizational climate", Litwin and Stringer's (1974) "Organizational Climate Scale" composed of 3 dimensions with totally 24 items was used. For the aim of this study, 2 dimensions with totally 17 items were utilized. The two dimensions used in the current study were named as; Structural Support and Organizational Recognition. Structural support dimension was composed of 10 items and Recognition dimension had 7 items. The Cronbach's alpha for "perceived organizational climate" construct was 0.91. For measuring "intrapreneurial behavior", "Intrapreneurial Behavior Scale" is adopted from Heinonen and Korvela (2003), which uses 39 items to measure 7 dimensions (Encouragement, individual motivation, transparency, openness and communality, individual competence, enabling working environment, encouragement to innovations, development). The Cronbach's alpha for "intrapreneurial behavior" construct was 0.88. "Organizational identification" measure was evaluated by a 25-item scale developed by Cheney (1983). The internal reliability (Cronbach α) is calculated as ,89 for organizational identification construct. All items were translated via a procedure of double-back translation and during the pretesting process; questionnaire items were found comprehensive. Overall, 81 items using 5 likert-type scale (ranging from 1=strongly disagree to 5=strongly agree) were used to measure 3 constructs of the research model.

3.4. Descriptive Results

The sample comprised of 210 employees working in 5 high performing and innovative firms operating in White Good Manufacturing, Food and Drink, Telecommunication, and Textile industries. 71% of the respondents were male, 72.2% were married; 72% were between ages 25–45. Majority of the samples (61.6%) had a university degree and 33.5% had occupational experience of 5-10 years. 81% of the respondents were working for 1–12 years in their current organization.

3.5. Reliability

Internal consistency of scales is calculated by Cronbach Alpha value. P-value (i.e. Sig. value) which indicates the likelihood of a particular outcome by chance, less than 5% were not considered statistically significant. Reliability analysis of the values of all the measuring instruments is greater than 0.80 as presented on Table 1.

Table 1. The summary statistics of survey (N=210)

Factors	N	Mean	α
Structural Support	10	3.78	0.90
Organizational Recognition	7	4.09	0.91
Organizational Climate	**17**	.....	**0.91**
Encouragement	16	3.45	0.89
Individual Motivation	5	3.92	0.88
Transparency&Openness&Communality	5	3.71	0.84
Individual Competence	4	4.02	0.91
Enabling Work Environment	3	4.05	0.82
Encouragement to Innovation	3	3.77	0.85
Development	3	3.88	0.89
Intrapreneurial Behavior	**39**	.....	**0.88**
Identification	15	3.72	0.86
Integrity	10	4.08	0.91
Organizational Identification	**25**	.....	**0.89**

3.6. Analyses and Results

It was revealed that all variables of the research model were significantly related to each other (Table 2). "Structural support" dimension of organizational climate had positive significant relationship with organizational identification (r=0,585; p<.01).

Table 2. Correlation Analysis of Variables (Pearson Correlation Analysis)

Variables:	1	2	3	4
1.Structural Support	1	0.423*	0,585*	0.465*
2.Organizational Recognition	0,423*	1	0,545*	0.418*
3.Organizational Identification	0,585*	0,545*	1	0,592*
4.Intrapreneurial Behavior	0.465*	0.418*	0.592*	1

Furthermore, the theoretical model of the study was expressed as organizational identification has a mediating role between organizational climate and intrapreneurial behaviors construct. To test this relationship, multiple regression analysis was conducted. In order to label organizational identification as a mediator between organizational climate and intrapreneurial behaviors, the following assessments were performed. Initially, independent variable's (structural support and organizational recognition) effect on the suggested mediating variable, organizational identification, (Hypothesis 2) was examined. Next, the suggested mediating variable's (organizational identification) significant effect on dependent variable (intrapreneurial behavior) (Hypothesis 3) was analyzed and independent variable's (organizational climate) significant effect on dependent variable (intrapreneurial behavior) (Hypothesis 1) was analyzed. In the final step, with the inclusion of the suggested mediating variable in the analysis, the effect of independent variable on dependent variable was tested (Hypothesis 4). According to the the three steps, the results of the regression analysis can be interpreted from the Table 3.

Table 3. Results of Regression Analysis

Independent Variables	Dependent Variables		
	1. Step	2. Step	3. Step
	Organizational Identification	Intrapreneurial Behaviors	Intrapreneurial Behaviors
Organizational Climate 1 (Structural Support)	0.427*	---	0.272*
Organizational Climate 2 (Organizational Recognition)	0.389*	---	0.783*
Organizational Identification	----	0.748*	0.683*
F-Value	36.574	173.125	45.185
R^2	0.454	0.542	0.571

p < 0.05

The first step in the model showed that organizational identification was taken as a dependent variable and organizational climate dimensions were taken as independent variables and regression analysis was applied. The results indicated that there was a significant and positive relationship between organizational identification and structural support and organizational recognition ($\beta 1 = 0.427$, $\beta 2 = 0.389$, $p < 0.05$). Thus, Hypothesis 3 was fully confirmed. Significant effect of both structural support and organizational recognition was found on intrapreneurial behavior ($\beta 1 = 0.463$, $\beta 2 = 0.822$,p

In the final step of the model, it was expected that in case of inclusion of the suggested mediating variable in the analysis, the effect of the independent variable on the dependent variable must be lower when the mediating variable is not included. With the inclusion of organizational identification in the analysis, the effect of the structural support on intrapreneurial behavior declined ($\beta = 0.272$, $p < 0.05$) with respect to the case while organizational identification was not in the analysis ($\beta = 0.463$, $p < 0.05$). The same results were found for organizational recognition ($\beta = 0.783$, $p < 0.05$). With the inclusion of the organizational identification, the effect of organizational recognition on intrapreneurial behavior declined and organizational identification was significant with respect to the case while organizational identification was not in the analysis ($\beta = 0.822$, $p < 0.05$). Therefore, the final step of the model (Hypothesis 4) was confirmed for explaining intrapreneurial behavior. In accordance with the multiple regression analyses results, the final research model is presented with Figure 2.

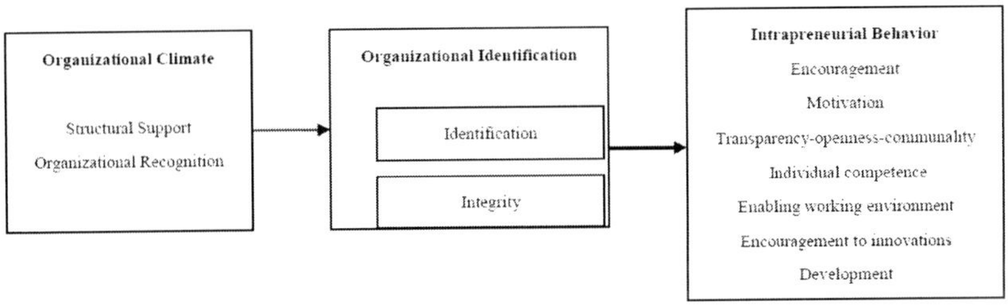

Figure 1. Final Research Model

4. CONCLUSION

The results of the current study have revealed that all variables of the research model had significant relationships between each other. The regression analyses reported that organizational climate dimensions of structural support and organizational recognition had significant positive effects on both organizational identification and intrapreneurial behavior construct. These findings supported the previous literature evidences which have indicated that organizational climate and organizational identification of employees had association with employee outcomes of intrapreneurial behaviors (e.g. Schollhammer, 1982; Kanter, 1984; Heinonen and Korvela, 2003; Antoncic and Hisrich, 2004; Antoncic and Zorn, 2004; Fitzsimmons et al, 2005; Antoncic, 2007; Vondey, 2010; Ergülen, 2011). Thus, the current study confimed that organizational climate is regarded as a sense of belonging for employees and the characteristics of the structural support and recognition served by the organization which are internalized by individuals are related to a employee behavioral outcomes intrapreneurial behaviors. This result can be supported with the implications of "social exchange theory", "social identity theory" and "organizational climate typology" since employees show intrapreneurial behaviors in accordance with their identification with the organization and how much they are structurally supported and recognized by the organization. In this study, establishing an intrapreneurial behavior in the organizations were viewed from a strategic management point of view since it has been addressed that intrepreneural behavior has cruical role in improving organizational competitiveness and enhancing growth. As known, today's managers have to face rapidly changing and fast-paced competitive environments and they are challenged to manage a wide variety of uncertainties. As Guth and Ginsberg (1990) indicated in 1990s, to cope with such challenges, the firms are increasingly turning to employee intrapreneurship as a means of growth and strategic orientation. Adopting intrapreneurship for establishing a long turn intrapreneurial orientation and strategic implementation has been also considered as a strategic choice (Jacobs and Kruger, 2001). Therefore, we suggest that enhancing intrapreneurial behaviors via supportive and innovative organizational climates should be seen as a crucial process among strategic planning approaches of the organizations. As such, this study integrated theory and practice from the fields of management, organizational behavior, strategic management and intrapreneurship.

However, as a limitation of this study, the current survey was performed among the employees working as lower and middle level managers and as non-managers of 5 innovative firms operating in White Good Manufacturing, Food and Drink, Telecommunication, and Textile industries in Turkey and which were selected from the official report of Turkish Time Journal (2011). Moreover, this study focused on the employees' perspective on each of the variables of the model. Both identification and intrapreneurial behavior constructs were measured through self-report method. It is suggested that future studies can be conducted within larger samples in different sectors and different size organizations for the reliability of the findings. Moreover, it is recommended that supervisor-report method or multiple source method for measuring intrapreneurial behavior can be used in future studies. Finally, it is suggested to investigate other potential determinants and antecedents of intrapreneurial behavior such as personality factors, environmental factors, organizational factors, leadership factors etc..

REFERENCES

1. Alpkan, L., Bulut, C., Gunday, G., Ulusoy, G., and Kilic, K. (2010), Organizational support for intrapreneurship and its interaction with human capital to enhance innovative performance, Management Decision, 48(5), pp.732-755.

2. Antoncic, B. (2007), Intrapreneurship: A comparative structural equation modeling study, Industrial Management and Data Systems, 107(3), pp.309-325.

3. Antoncic, B. and Hisrich, R. D. (2001). Intrapreneurship: Construct refinement and cross-cultural validation. Journal of Business Venturing, 16(5), pp.495-527.

4. Antoncic, B. and Hisrich, R. D. (2003), Clarifying the intrapreneurship concept, Journal of small business and enterprise development, 10(1), pp.7- 24

5. Antoncic, B. and Hisrich, R. D. (2004), Corporate entrepreneurship contingencies and organizational wealth creation, Journal of Management Development, 23(6), pp.518-550.

6. Antoncic, B. and Zorn, O. (2004), The mediating role of corporate entrepreneurship in the organizational support-performance relationship: An empirical examination, Managing Global Transitions, 2(1), pp.5-14.

7. Antoncic, J. A. and Antoncic, B. (2011), Employee satisfaction, intrapreneurship and firm growth: a model, Industrial Management & Data Systems, 111(4), pp.589-607.

8. Ashforth, B. E. and Mael, F. (1989), Social identity theory and the organization, Academy of Management Review, 14(1), pp.20-39.

9. Badguerahanian, L. and Abetti, P. A. (1995), The rise and fall of the Merlin-Gerin Foundry Business: A case study in French corporate entrepreneurship, Journal of Business Venturing, 10(6), pp.477-493.

10. Bellou, V. and Andronikidis, A. I. (2009), Examining organizational climate in Greek hotels from a service quality perspective, International Journal of Contemporary Hospitality Management, 21(3), pp.294-307

11. Bindl, U.K. and Parker, S.K. (2011), Proactive work behavior: Forward-thinking and change- oriented action in organizations, Unpublished Doctorate Thesis, Institute of Work Psychology, University of Sheffield, United Kingdom.

12. Carrier, C. (1996), Intrapreneurship in small businesses: An exploratory study, Entrepreneurship Theory and Practice, Fall, pp.5-20

13. Cheney, G. (1983), The rhetoric of identification and the study of organizational communication, Quarterly Journal of Speech, 69(2), pp.143-158

14. Cole, M. S. and Bruch, H. (2006), Organizational identity strength, identification, and commitment and their relationships to turnover intention: Does organizational hierarchy matter?, Journal of Organizational Behavior, 27(5), pp.585-605.

15. Covin, J.G. and Slevin, D.P. (1991), A conceptual model of entrepreneurship as firm behavior, Entrepreneurship Theory & Practice, 16(1), pp. 7-25

16. Cunningham, J. B. and Lischeron, J. (1991). Defining entrepreneurship. Journal of small business management, 29(1), pp.45-61

17. Dukerich J.M.,Golden, B.,and Shortell S.M.(2002), Beauty is in the eye of the beholder: The impact of organizational identification, identity and image on physician cooperative behavior, Administrative Science Quarterly, 47(3), 1-31.

18. Ergülen, G.K. (2011), Influence of leader-member exchange quality on organizational attachment around the organizational climate perspective in family-owned business. Unpublished Master Thesis, Marmara University, Social Sciences Institute, Department of Organizational Behavior, Istanbul.

19. Evidence, Organization Studies, 8(1), pp.39-59

20. Farooqui, M. R. (2012), Measuring organizational citizenship behavior (OCB) as a consequence of organizational climate (OC), Asian Journal of Business Management, 4(3), pp.294-302.

21. Fitzsimmons, J. R., Dougles, E. J., Antoncic, B., and Hisrich, R. D. (2005), Intrapreneurship in Australian firms. Journal of the Australian and New Zealand Academy of Management, 11(1), pp.17–27.

22. Friedlander, F. and Greenberg, S. (1971), Effect of job attitudes, training, and organization climate on performance of the hard-core unemployed. Journal of Applied Psychology, 55(4), pp.287-311

23. Guth, W.D. and Ginsberg, A. (1990), Guest editor's introduction: Corporate entrepreneurship. Strategic Management Journal (special issue), 11(5), pp.5–15

24. Harris,G.E. and Cameron,J.E. (2005), Multiple dimensions of organizational identification, commitment as predictors of turnover intentions&psychological well-being, Canadian Journal of Behavioural Science, 37(3), pp.135-159.

25. Hatch, M. J. and Schultz, M. (1997), Relations between organizational culture, identity and image, European Journal of Marketing, 31(5/6), pp.356- 365.

26. Heinonen, J. (1999), Towards Customer Orientation and Competitiveness. The Potential of Intrapreneurship in the Change Process of a Municipal Service Unit. Publications of the Turku School of Economics and Business Administration Series A-5.

27. Heinonen, J. and Korvela, K. (2003). How about measuring intrapreneurship?. In Conference Proceedings of 33rd EISB (Entrepreneurship, Innovation and Small Business). Conference in Milan, Italy

28. Holloway, J. B. (2012), Leadership behavior and organizational climate: An empirical study in a non-profit organizational, Emerging Leadership Journeys, 5(1), pp.9-35.

29. Jacobs, H. and Kruger, S. (2001). Establishing an intrapreneurial orientation as strategy: a framework for implementation. Acta Commercii, 1(1), pp.1-11.

30. Jones, G. R. and Butler, J. E. (1992), Managing internal corporate entrepreneurship: An agency theory perspective, Journal of Management, 18(4), pp.733-749

31. Kanten, P. and Ülker, F. E. (2013), The effect of organizational climate on counterproductive behaviors: An empirical study on the employees of manufacturing enterprises, The Macrotheme Review, 2(4), Summer, pp.144-160.

32. Kanter, R. M. (1984), Change masters. Simon and Schuster Publishing Hause.

33. Khandwalla, P.N. (1987), Generators of pioneering-innovative management: Some Indian

34. Kreiner, G. E. and Ashforth, B. E. (2004), Evidence toward an expanded model of organizational identification. Journal of Organizational Behavior, 25(1), pp.1-27.

35. Lee, H. J. (2004), The role of competence-based trust and organizational identification in continuous improvement, Journal of Managerial Psychology,19(6), pp.623-639.

36. Litwin,G.H. and Stringer,R.A.(1974),Motivation&Organizational Climate, 4rd Ed., Harvard University,Boston, 1974.

37. Mael, F. A. and Ashforth, B. E. (1995), Loyal from day one: Biodata, organizational identification, and turnover among newcomers, Personnel Psychology, 48(2), pp.309-333.

38. Mael, F. A. and Ashforth, B. E. (2001), Identification in work, war, sports, and religion: Contrasting the benefits and risks, Journal for the Theory of Social Behaviour, 31(2), pp.197-222.

39. Miller, D. (1983), The correlates of entrepreneurship in three types of firms, Management Science, 29, pp.770-791.

40. Miller, V. D., Allen, M., Casey, M. K., and Johnson, J. R. (2000), Reconsidering the organizational identification questionnaire, Management Communication Quarterly, 13(4), pp.626-658.

41. Moghimi, S. and Subramaniam, I. D. (2013), Employees' creative behavior: The role of organizational climate in Malaysian SMEs, International Journal of Business & Management, 8(5), pp.1-13.

42. Özdemir, F. (2006), Örgütsel iklimin iş tatmin düzeyine etkisi, Unpublished Doctorate Thesis, Çukurova University, Social Sciences Institute, Adana.

43. Pinchot, G. (1985), Intrapreneuring: Why you don't have to leave the corporation to become an entrepreneur. University of Illinois at UrbanaChampaign's Academy for Entrepreneurial Leadership Historical Research Reference in Entrepreneurship

44. Rahimic, Z. (2013), Influence of organizational climate on job satisfaction in Bosnia and Herzegovina companies, International Business Research, 6(3), 1-18.

45. Resick, C. J., Hanges, P. J., Dickson, M. W., and Mitchelson, J. K. (2006), A cross-cultural examination of the endorsement of ethical leadership, Journal of Business Ethics, 63(4), pp.345-359.

46. Riketta, M. (2005), Organizational identification: A meta-analysis, Journal of Vocational Behavior, 66(2), pp.358-384.

47. Schollhammer, H. (1982), Internal corporate entrepreneurship C.A. Kent, D.L. Sexton, K.H. Vesper (Eds.), Encyclopedia of EntrepreneurshipPrentice Hall, Englewood Cliffs, NJ (1982), pp. 209–223

48. Schrodt, P. (2002), The relationship between organizational identification and organizational culture: Employee perceptions of culture and identification in a retail sales organization, Communication Studies, 53(2), pp.189-202.

49. Stevenson, H. H. and Jarillo, J. C. (1990), A paradigm of entrepreneurship: entrepreneurial management, Strategic Management Journal, 11(5), pp.17-27.

50. Stopford, J. M. and BadenǦ Fuller, C. W. (1994), Creating corporate entrepreneurship, Strategic Management Journal, 15(7), pp.521-536

51. Tajfel, H. and Turner, J.C. (1985), The social identity theory of intergroup behaviour, In Worchel, S. and Austin, W.G. (Eds), Social Psychology of Intergroup Relations, 2nd Edition, Nelson-Hall, Chicago, IL, pp.7-24.

52. Tajfel, H. E. (1978), Differentiation between social groups, Academic Press

53. Taştan, S.B. (2012), Bir pozitif psikoloji kavramı olarak örgütle özdeşleşmenin psikolojik güçlendirme algısı ve gönüllü performans davranışı arasındaki ilişkide ara değişken rolünün değerlendirilmesi: Gıda sektöründe yapılan bir araştırma, Organization and Management Sciences Journal, 4(1), pp.227-238.

54. Tüzün, İ. K. and Çağlar, İ. (2009), Investigating the antecedents of organizational identification, Journal of Doğuş University, 10 (2), pp.284-293. Turkish Time Journal (2011), The most innovative firms in Turkey, 116, December, pp.78-79. Van der Sijde, P., Veenker, S., and During, W. (2013), Intrapreneurship in SMEs: About the role of management and R&D, European Journal of Business and Social Sciences, 1(11), pp.24-30.

55. Vesper, K. H. (1990), New venture strategies. University of Illinois at Urbana-Champaign's Academy for Entrepreneurial Leadership Historical Research Reference in Entrepreneurship.

56. Vondey, M. (2010), The relationships among servant leadership, organizational citizenship behavior, person-organization fit, and organizational identification, International Journal of Leadership Studies, 6(1), pp.3-27.

57. West, M., Shackleton, V., Dawson, J. F., Lawthom, R., Maitlis, S., Robinson, D. L., and Wallace, A. M. (2004), Development & validation of an organizational climate measure, Aston Business University

58. Zahra, S. A. (1996), Technology strategy and new venture performance: a study of corporate-sponsored and independent biotechnology ventures, Journal of Business venturing, 11(4), pp.289-321.

59. Zahra, S.A. (1993), Environment, corporate entrepreneurship, and financial performance: A taxonomic approach, Journal of Business Venturing, 8(4), pp.319-340.

60. Zhang, J. and Liu, Y. (2010), Organizational climate and its effects on organizational variables: An empirical study, International Journal of Psychological Studies, 2(2), 1-26.

Chapter 12

Employee Compensation: Theory, Practice, and Evidence

Barry A. Gerhart, Harvey B. Minkoff, Ray N. Olsen

Center for Advanced Human Resource Studies Cornell University TRW Corporation

ABSTRACT

As organizations continue to face mounting competitive pressures, they seek to do more with less and do it with better quality. As goals for sales volume, profits, innovation, and quality are raised, employment growth is often tightly controlled and in many cases, substantial cuts in employment have been made. To accomplish more with fewer employees calls for effective management of human resources. Typically, the employee compensation system, the focus of this chapter, plays a major role in efforts to manage human resources better.

Keywords: employee, compensation, organization, profit, human, resource, manage, pay, market

INTRODUCTION

As organizations continue to face mounting competitive pressures, they seek to do more with less and do it with better quality. As goals for sales volume, profits, innovation, and quality are raised, employment growth is often tightly controlled and in many cases, substantial cuts in employment have been made. To accomplish more with fewer employees calls for effective management of human resources. Typically, the employee compensation system, the focus of this chapter, plays a major role in efforts to manage human resources better

Employee compensation plays such a key role because it is at the heart of the employment relationship, being of critical importance to both employees and employers. Employees typically depend on wages, salaries, and so forth to provide a large share of their income and on benefits to provide income and health security. For employers, compensation decisions influence their cost of doing business and thus, their ability to sell at a competitive price in the product market. In addition, compensation decisions influence the employer's ability to compete for employees in the labor market (attract and retain), as well as their attitudes and behaviors while with the employer.

Employee compensation practices differ across employment units (e.g., organizations, business units, and facilities) on several dimensions (Gerhart & Milkovich, 1990, 1992; Gerhart, Milkovich, & Murray, 1992). The focus of the employee compensation literature has been on defining these dimensions, understanding why organizations differ on them (determinants), and assessing whether such differences have consequences for employee attitudes and behaviors, and for organizational effectiveness. In the following discussion, we briefly describe the basic dimensions of compensation and summarize some of the key theories used to explain the consequences of different compensation decisions. A discussion of pay determinants can be found in Gerhart and Milkovich (1990, 1992).

Strategic Pay Dimensions

Pay practices vary significantly across employing units and to some degree, across jobs. We discuss the form, level, structure, mix, and administration of payment systems (Gerhart & Milkovich, 1992; Heneman & Schwab, 1979; Milkovich & Newman, 1993).

First, pay can be in the form of cash or benefits (e.g., health care, retirement, paid vacation). On average, about 70 percent of payments to U.S. employees are in the form of cash, leaving 30 percent in the form of noncash and deferred cash benefits (Noe, Hollenbeck, Gerhart, & Wright, 1994). Health care has been the fastest growing benefit, and most employers describe the challenge of controlling this cost while providing quality coverage as one of their top human resource management challenges.

Second, both benefits and cash compensation can be described in terms of their level (how much). Most organizations use one or more market pay surveys to help determine what other organizations pay specific jobs in making their own pay level decisions. More broadly, total labor costs are a function of both compensation cost per employee and total employee headcount. Therefore, to assess competitiveness in the product market, organizations should not focus only on pay levels. They should compare total labor costs, and better yet, they should compare with other organizations the sort of return (or productivity) they receive in terms of profits, sales, and so forth for each dollar spent on labor costs. The now common announcements of major reductions in force attest to the importance of controlling labor costs. Such decisions are also sometimes driven by comparisons of revenue or profits per employee, or the ratio of sales or profits to labor costs.

Labor costs and productivity are also key factors in decisions about where to locate production. Germany's high labor costs have led to what Business Week described as the "Exodus of German Industry." German companies are moving production to lower labor cost countries, such as those in Eastern Europe and the United States.1 BMW recently announced it would be building vehicles in South Carolina and Mercedes-Benz will produce vehicles in Alabama. Agreements such as the North American Free Trade Agreement.(NAFTA) and the General Agreement on Trade and Tariffs will only reinforce the globalization of production. However, contrary to what was heard in the debate over NAFTA, labor costs will not be the determining factor in most cases, except perhaps for labor-intensive production. Labor costs as a percentage of total costs is shrinking in many cases, and other factors such as access to markets and labor force quality will often be more important. The decision by BMW and Mercedes-Benz to build in the United States, not Mexico, is evidence of this.

Third, the structure refers to the nature of pay differentials within an employing unit. How many steps or grades are in the structure? How big are the pay differentials between different levels in the structure? Large organizations often have over 20 such levels, although many organizations have recently reduced the number of steps ("delayered"). Are employees at the same hierarchical level in different parts of the organization (e.g., different product sectors or different occupational groups) paid the same? Yet another aspect of structure is the timing of payment over employees' careers. Some organizations may bring entry level people in at a relatively high rate of pay, but then provide relatively slow pay growth, while another organization may bring employees in relatively low but offer greater opportunities for promotion and pay growth over time.

Fourth, payment systems differ in their mix (how and when cash compensation is disbursed). Some organizations pay virtually all employees a base salary that is adjusted approximately once per year through a traditional merit increase program. Merit increases become part of base salary and are supposed to depend on merit (performance), although there is a widespread belief that most employees get about the same percentage increase, regardless of their performance. As described below, an increasing number of organizations are using so-called variable pay or pay at risk, which means that some portion of employees' pay is uncertain and depends on some combination of future business unit or organization performance (e.g., profits, stock performance, productivity), group performance, and individual performance. Specific pay programs that influence pay mix are merit pay, incentive pay, gainsharing, profit sharing, and stock plans (e.g., stock options).

Fifth, pay is administered differently in different organizations. The design of pay policies differs, for example, in terms of who is involved in the process. The roles of human resource departments, line managers, and rank and file employees differ across situations. In some organizations, line managers may design plans, often with assistance from the human resources department. Alternatively, human resources takes the lead in other cases. Employees to be covered by a payment system are sometimes involved, and in some cases, may actually design plans for themselves.

Communication is another aspect of administration. The most technically sophisticated payment plan can generate desired employee reactions or exactly the opposite. The actual effect depends on whether the rationale for the payment plan is understood and accepted and whether employees' perceptions of the facts upon which the rationale is built (e.g., the company's financial health, the pay of employees in other jobs or organizations) are the same as the perceptions of those charged with seeing that the payment plan has the intended effects.

We focus in this chapter on cash compensation issues. Benefits warrants a chapter of its own and discussions are available elsewhere (Beam & McFadden, 1992; Gerhart & Milkovich, 1992; Noe, Hollenbeck, Gerhart, & Wright, 1994). Further, our discussion of cash compensation is mostly limited to pay mix issues, an area that has been of great interest to organizations as they move (or consider a move) to "new" programs such as stock plans for non-executives, gainsharing, and profit sharing. In the remainder of this chapter, we provide a survey of theories that have been used to study the effects of pay decisions, describe specific pay programs and their expected consequences, and review recent empirical evidence on that question.

Consequences Of Pay Decisions: Theories

To understand what types of pay systems are most likely to be effective and how their effectiveness differs according to contingency factors such as business strategy, national culture, competitive environment, and employee characteristics, we need to have a good conceptual framework, or theory. In truth, there is as of yet no grand theory of compensation that takes these contingency factors into account, although recent work by Gomez-Mejia and Balkin (1992) is promising.

In examining consequences, we need to recognize that effectiveness is a multi-faceted concept that could include at a minimum, cost, productivity, innovation, quality, financial, and attitudinal dimensions. Further, the relative importance of these dimensions will vary across organizations and business units

At the individual level of analysis, theories have been used to show how pay plans can be used to energize, direct, and control employee behavior. We briefly describe three such theories used in research on pay

Reinforcement and Expectancy Theories

Reinforcement theory states that a response followed by a reward is more likely to recur in the future (Thorndike's Law of Effect). The implication for compensation management is that high employee performance followed by a monetary reward will make future high performance more likely. By the same token, high performance not followed by a reward will make it less likely in the future. The theory emphasizes the importance of a person actually experiencing the reward.

Like reinforcement theory, expectancy theory (Vroom, 1964) focuses on the link between rewards and behaviors (instrumentality perceptions), although it emphasizes expected (rather than experienced) rewards (i.e., incentives). Motivation is also a function of two other factors: expectancy, the perceived link between effort and performance, and valence, the expected value of outcomes (e.g., rewards). Compensation systems differ according to their impact on these motivational components. Generally speaking, pay systems differ most in their impact on instrumentality: the perceived link between behaviors and pay, also referred to in the pay literature as "line of sight." Valence of pay outcomes should remain the same under different pay systems. Expectancy perceptions often have more to do with job design and training than pay systems.

Equity Theory

Equity theory suggests that employee perceptions of what they contribute to the organization, what they get in return, and how their return-contribution ratio compares to others inside and outside the organization,' determine how fair they perceive their employment relationship to be (Adams, 1963). Perceptions of inequity are expected to cause employees to take actions to restore equity. Unfortunately, some such actions (e.g., quitting or lack of cooperation) may not be helpful to the organization.

Two recent empirical studies provide good examples of the types of counterproductive behaviors that can occur as a result of perceived inequity. In the first study, Greenberg (1990) examined how an organization2 communicated pay cuts to its employees and the effects on theft rates and perceived equity. Two organization units received 15% across-the-board pay cuts. A third unit received no pay cut and served as a control group. The reasons for the pay cuts were communicated in different ways to the two pay-cut groups. In the "adequate explanation" pay-cut group, management provided a significant degree of information to explain its reasons for the pay cut, and also expressed significant remorse. In contrast, the "inadequate explanation" group received much less information and no indication of remorse. The control group received no pay cut (and thus no explanation).

The control group and the two pay-cut groups began with the same theft rates and equity perceptions. After the pay cut, the theft rate was 54% higher in the adequate explanation group than in the control group. However, in the "inadequate explanation" condition, the theft rate was 141% than in the control group. In this case, communication had a large, independent effect on employees' attitudes and behaviors.

Cowherd and Levine (1992) used a sample 102 business units in 41 corporations to examine whether the size of the pay differential between lower-level employees and top management had any impact on product quality. Cowherd and Levine suggest that individuals often compare their pay to that of people higher in the organization structure. If lower-level employees feel inequitably treated, they may seek to reduce their effort to achieve equity. Quality, in their study, was defined as customer perceptions of the quality of goods and services. They hypothesized that extrarole, or citizenship behaviors, such as freely offering to help others, following the spirit rather than letter of rules, and correcting errors that would ordinarily escape notice, would be less likely when pay differentials between hourly and top managerial employees were large. Their results supported this hypothesis, suggesting that organizations need to take care that they not forget the potential adverse motivational consequences of executive pay for the motivation of other employees.

Agency Theory

Agency theory, until recently best known in the economics, finance, and law literatures, focuses on the divergent interests and goals of the organization's stakeholders, and the ways that employee compensation can be used to align these interests and goals (Eisenhardt, 1989; Fama & Jensen, 1983). ownership and management (or control) are typically separate in the modern corporation, unlike the days when the owner and manager were often the same person. With most stockholders far removed from day-to-day operations, so-called agency costs (i.e., costs that arise from the interests of the principals/owners and their agents/managers not converging are created. What is best for the agent/manager, may not be best for the owner.

Examples of agency costs include management spending money on perquisites (e.g., "superfluous" corporate jets) or "empire building" (acquisitions that do not add value to the company but may enhance the manager's prestige or pay) rather than seeking to maximize shareholder wealth (Lambert & Larcker, 1989). In addition, the fact that managers and shareholders may differ in their attitudes toward risk gives rise to agency costs. Shareholders can diversify their investments (and thus their risks) more easily than managers can diversify risk in their pay. As a consequence, managers may prefer relatively little risk in their pay (e.g., high emphasis on base salary,-low emphasis on uncertain bonuses or incentives). Indeed, research shows that managerial compensation in manager-controlled firths is more often designed in this manner (Tosi & Gomez-Mejia, 1989). Agency costs also stem from differences in decision-making horizons. Especially where managers expect to spend little time in the job or with the organization, they may be more inclined to maximize short-run performance (and pay), perhaps at the expense of long-term success.

Agency theory is also of value in the analysis and design of non-managers' compensation. In this case, the divergence of interests may exist between managers (now in the role of principals) and their employees (who take on the role of agents). In designing either managerial or non-managerial compensation, the key question is, "How can such agency costs be minimized?" Agency theory says that the principal must choose a contracting scheme that helps align the interests of the agent with the principal's own interests (i.e., reduces agency costs). These contracts can be classified as either behavior oriented (e.g., merit pay) or outcome oriented (e.g., stock options, profit sharing, commissions).

At first blush, outcome-oriented contracts seem to be the obvious solution. If profits are high, compensation goes up. If profits go down, compensation goes down. The interests of "the firm" and employees are aligned. An important drawback, however, is that such contracts increase the amount of risk borne by the agent. Furthermore, because agents are averse to risk, they may require higher pay (a compensating wage differential) to make up for it.

Behavior-based contracts, on the other hand, do not transfer risk to the agent, and thus do not require a compensating wage differential. However, the principal must be able to monitor with little cost what the agent has done. Otherwise, the principal must either invest in monitoring/information or structure the contract so that pay is linked at least partly to outcomes. Which type of contract should an organization use? It depends partly on the following factors (Eisenhardt, 1989):

- Risk aversion. Risk aversion among agents makes outcome-oriented contracts more costly.
- Outcome uncertainty. Profit is an example of an outcome. Linking pay to profits (outcome-based contract) is more costly to the extent that profits vary and so there is a risk of low profits.
- Job programmability. As jobs become less programmable (i.e., less routine and less structured), and more difficult to monitor, outcome-oriented contracts become more likely. The increasing complexity of organizations and technology makes monitoring more difficult, and may help explain the growing use of variable pay programs (discussed below), which are examples of outcome-based contracts. Consistent with this idea, outcome-oriented contracts (e.g., profit sharing and stock plans) are more prevalent in research and development organizations, where monitoring is especially difficult (Milkovich, Gerhart, & Hannon, 1991). Pay levels are also higher, consistent with the idea that employees must be compensated for sharing more risk.
- Measurable lob outcomes. When outcomes are more measurable, outcome-oriented contracts are more likely.
- Ability to pay. outcome-oriented contracts contribute to higher compensation costs because of the risk premium.
- Tradition. A tradition or custom of using (or not using) outcome-oriented contracts will make such contracts more (or less) likely.

Influences on Labor Force Composition

Traditionally, the theories described above have been used to understand how using pay to recognize individual contributions can influence the behaviors and attitudes of current employees, whereas pay level and benefits have been seen as a way to influence so-called membership behaviors: decisions about whether to join or remain with the organization.

However, there is increasing recognition that individual pay programs may also have an effect on the nature and composition of an organization's work force (Milkovich & Wigdor, 1991; Gerhart & Milkovich, 1992). For example, it is possible that an organization that links pay to performance may attract more high performers than an organization that does not link the two. There may be a similar effect with respect to job retention.

Breaking things down further, perhaps organizations that link pay to individual performance are more likely to attract individualistic types of employees, while organizations relying more heavily on team rewards are more likely to attract more team-oriented employees. Although there is no concrete evidence of this yet, it has been found that different pay systems attract different people depending on their personality traits and values (Bretz, Ash, & Dreher, 1989; Judge & Bretz, 1992). The implication is that the design of compensation programs needs to be carefully coordinated with the business and human resource strategy.

STRATEGY

Moving from the individual level of analysis to the business unit and corporate level, there are theories of what corporate and pay strategies fit best together. Stage in the product life cycle (Ellig, 1981) and the degree and process of diversification (Kerr, 1985) have been raised as contingency factors in the design of pay strategies (Milkovich, 1988). Briefly, organizations (or probably more precisely, business units) may go through growth, maintenance, and decline stages, each of which calls for a different compensation strategy. For example, in the growth stage, it was recommended that there be substantial pay at risk to provide high upside earnings potential (e.g., using stock plans) to spur innovation, growth, and risk-taking, combined with low fixed costs (base salary and benefits) to preserve scarce capital for investment. In the maintenance and decline stages, there would be less emphasis on pay at risk (except perhaps for more short term focused plans), and more dollars allocated to base salary and benefits.

The literature on diversification and pay strategy suggests that single product firths and unrelated product firms (e.g., conglomerates) have more pay at risk than related product firms, and pay is more decentralized and tied to business unit rather than corporate performance in the unrelated product firms. This flexibility makes sense where each business unit has independent goals, and there is little need for coordination and thus for consistency in pay practices. From an agency theory point of view, it may be more necessary to rely on outcome-oriented contracts in unrelated products firms because the market-specific expertise is concentrated in the business units, making it difficult for corporate headquarters to make evaluations using behavior-oriented contracts. Finally, an unrelated firm that is a result of mergers and acquisitions is more likely to have the flexibility and pay linked to unit performance than an unrelated firm that is the result of internal growth because there is often more interdependence and interaction in the latter case (Kerr, 1985)

Gomez-Mejia and Balkin (Gomez-Mejia & Balkin, 1992; Gomez-Mejia, 1992) have summarized much of the research on these questions, and provided some of the first tests of whether firms that choose pay strategies consistent with the above frameworks actually perform better. The answer seems to be "yes." They have provided propositions about which types of pay practices are likely to be most effective based on various strategy frameworks. For example, the Miles and Snow (1978) model classifies business units as defenders (stable markets, focus on efficiency), prospectors (focus on new markets and technologies), and analyzers, which have elements of both defenders and prospectors. According to Gomez-Mejia and Balkin, variable pay, for example, should be higher in the prospector business units than in the defender business units. Table 1 shows other proposed differences.

Table 1. Matching Organization Strategy and Pay Strategy

	Business Unit Strategy	
	Defenders	*Prospectors*
Pay Strategy Dimensions		
Risk Sharing (Variable Pay)	Low	High
Time Orientation	Short-term	Long-term
Pay Level (short run)	Above market	Below market
Pay Level (long run potential)	Below market	Above Market
Benefits Level	Above market	Below market
Centralization of Pay Decisions	Centralized	Decentralized
Pay Unit of Analysis	Job	Skills

Source: Adapted from Gomez-Mejia, L.R. & Balkin, D.B. (1992). Compensation, organizational strategy, and firm performance. Appendix 4b.

SUMMARY

Reinforcement, expectancy, and agency theories all focus on the fact that behavior-reward contingencies can shape behaviors. However, agency theory is of particular value in studying variable pay because of its emphasis on the risk-reward trade-off, an issue that needs close attention when considering variable pay plans, which can carry significant risk. Equity theory is also very relevant because it can be applied to just about any pay decision, because fairness is always a key concern.

Moving away from individual-level theories, life cycle and diversification-based contingency theories suggest that pay strategies should fit with corporate strategies. The evolving empirical literature provides tentative support for many of the specific propositions

PAY PROGRAMS

Table 2 summarizes the key features of some of the most widely used pay programs. Key dimensions include the payment method (whether increases roll into base salary or are paid as bonuses or equity), the frequency of payouts, the nature of the performance measure, and who is typically covered under the different plans.

Table 2. Comparison of Different Pay Programs

	Individual Incentives	Merit Pay	Merit Bonus	Gainshairing	Profit Sharing	Ownership	Skill Based Pay
Payment Method	Bonus	Changes in base pay	Bonus	Bonus	Bonus	Equity Changes	Changes in base pay
Payout Frequency	Weekly	Annually	Annually	Monthly or quarterly	Semi-annually or annually	When stock sold	When skill required
Performance Measurement	Output, productivity, sales	Performance rating	Performance rating	Production or controllable costs	Profit	Stock value	Skill acquisition
Coverage	Direct labor	All employees	All employees	Production or service unit	Total organization	Total organization	All employees

In compensating employees, an organization does not have to choose one program over another. Instead, a combination of programs is often the best solution. For example, one program may foster teamwork and cooperation but not enough individual initiative. Another may do the opposite. Used in conjunction, a balance may be attained. We now turn to a discussion of some recent trends in pay and an evaluation of .where such trends are likely to lead us.

RECENT DEVELOPMENTS

The Shift to Variable Pay

According to a survey of over 2,000 U.S. companies by Hewitt Associates (Tully, 1993), the percentage of companies having a variable pay policy covering all salaried employees increased from 47% in 1988 to 68 in 1993. Moreover, whereas the standard merit increase (which rolls into base salary) was larger (5% versus 3.9%) in 1988 than the merit bonus (a lump sum payment that does become part of base salary), by 1993 the situation was reversed with the merit bonus being larger on average than the standard merit increase (5.9% versus 4.3%).

Those in the human resource management field expect the movement toward variable pay to continue. In the Workplace 2000 study conducted by Dyer and Blancero (1993), 57 human resource executives, consultants, academics, and others were asked to describe how the workplace was likely to change by the year 2000. Dyer and Blancero provided study participants the characteristics of a hypothetical service organization in 1991 and asked how it would look in the year 2000. One expectation of participants was that pay would become more variable. As Table 3 indicates, variable pay as a percentage of total direct compensation was expected to increase significantly for each of the four occupational groups studied.

Table 4 provides some examples of how variable pay programs operate.

Table 3. Variable Pay as a Percentage of Total Direct Compensation 1991 Scenario and Year 2000 Projection

Occupational Group	1991	2000	Percentage Change
Executives	20 %	33 %	65 %
Managers	10 %	23 %	130 %
Professional/Technical	10 %	18 %	80 %
Support	10 %	14 %	40 %

Table 4. Examples of Variable Pay Programs for Managers

Company	Plan Participants	Base Pay Policy	Bonus Policy
Nucor Steel	14 plant managers	$80,000 to $150,000 (25 % below market)	5 % of every dollar earned beyond 10 % return on equity goes into bonus pool. Last year, average plant manager bonus equaled base salary
General Mills	Managers (Marketing manager in this example)	$75,000 (versus $90,000 market midpoint)	$10,000 if profits growth and return on capital are at market average; up to $40,500 if profits growth and return on capital are in top 10 % of market
AT&T	80,000 middle managers& 30,000 scientists, researchers, technical employees	Between 1986 and 1989, pay raises less than one-half of competitors--move from pay leader to below midpoint	1. Individual or team bonus--pool depends in part on corporate net profitability (5 to 15 % of base depending on individual/team performance) 2. Business unit net profitability (about 2 % of base or less this year) 3. Corporate net profitability (7 to 11 % of base)

Group and Organization-Based Variable Pay

Dyer and Blancero (1993) also found a strong belief that, in the future, variable pay would be based to a lesser degree on individual performance and to a much greater degree on firth, business unit, and work group performance (see Table 3). The examples in Table 4 are consistent with this expectation. It should be noted, however, that despite these significant changes, Dyer and Blancero found that individual performance is expected to remain as the single most important determinant of variable pay for all occupational groups

Why are organizations making greater use of variable pay, and why are they moving away from an individual focus to more of a group and organization focus? Variable pay is seen as a way of both controlling costs (especially in the case of organization-wide plans) and re-directing employee behavior.

Better cost control is expected to be gained by replacing standard merit increases with merit bonuses that are linked to firm or business unit performance. Thus, when profits or stock returns are good, they can be shared with employees. However, when profits or stock returns are poor or nonexistent, the organization is not saddled to the same degree with high fixed labor costs.

In theory, the use of variable pay plans to control labor costs is fine and it even works in practice under the right conditions; namely, if employees see a compelling business need to stay competitive in this manner. However, as in the widely discussed case of the DuPont Fibers division variable pay plan (Santora, 1991),3 employee opposition to downside variability in pay when profit targets are not met can lead to such plans being discontinued as soon as the labor cost control aspect is supposed to kick in, and employees forcgo bonuses and receive only their (below market) base salary. This result is consistent with agency theory's prediction that outcome-oriented contracts are less successful when there is high outcome uncertainty.

Somc organizations seek to avoid this "problem" by setting base pay at a higher level, and then sharing profits or stock with employees on top of their base salary during good years. These "gravy" plans do not control labor costs and, in fact, raise them. Yet, unless there is a compelling reason to believe that such pay plans significantly raise employee or for organization productivity, organizations following this approach run the risk of investing extra money in the form of labor costs without realizing any return on the investment. Therefore, consistent with agency theory, employees may demand a compensating pay premium to assume risk. So, pay risk costs the organization more money, but gains in effectiveness are not certain.

Organizations that use variable pay, with or without downside risk, often believe that such plans do generate significant returns. In agency theory terms, profit-sharing, stock plans, and gainsharing are examples of outcome-based contracts that seek to align the interests of employees and management with those of owners. As such, they are expected to re-direct behavior away from parochial individual goals, and more toward what it takes in terms of cooperation, commitment, and innovation to make the group, business unit, or organization a success

A change to variable pay may be a way to send a message to employees that things are going to change in important ways and therefore, may be helpful in supporting other major human resource changes. For example, variable pay may support a move to a team-based organization. As another example, variable pay may help eliminate the "entitlement" mentality or culture that can result from so-called merit increase plans that (in fact) fail to differentiate between employees with different performance levels, roll the increase into base salary so the cost remains in future years, and ignore the performance of the business. With a merit bonus, the pay has to be re-earned each year. Past individual performance does not matter, and is not reflected in base salary. Therefore, employees cannot rest on their past laurels. Moreover, the bonus pool may be linked to organization or business unit profitability. Again, the idea is to align employee interests with those of the organization. In this case, the goal is to encourage continuous improvement and a forward-looking perspective.

Agency theory suggests that group and organization incentives can also contribute to greater overall levels of performance monitoring by, in effect, making each employee a principal who monitors other employees (Levine & Tyson, 1990). So, if your pay and my pay depend on what we do as a team, we will be more likely to monitor each other's performance and give feedback to one another when performance needs improvement: Similarly, according to equity theory, if a person feels that his or her inputs (e.g., work effort) are greater than another member of the work group, but they receive the same reward, one way to restore equity would be to encourage (or pressure) the other person to put forth more effort

Group size, however, is a key contingency variable in discussions of the behavioral impact of group and organization variable pay plans. According to expectancy theory, the larger the number of employees covered by a pay plan, the weaker the link they see between their own performance and pay (Schwab, 1973), and thus the weaker is their motivation. Similarly, a theme from the shirking, social loafing, and free rider literatures is that individual effort decreases as the size of the group increases (Kidwell & Bennett, 1993).

The implication, therefore, is that the ability of group and organization plans to change employee behavior may be very limited in cases where large numbers of employees are covered. On this dimension, gainsharing plans, which typically cover smaller groups of employees, probably have an advantage over organization-wide plans like profit sharing and stock-based plans. Another advantage is that the performance measures in gainsharing plans (e.g., labor costs, quality) are often more controllable, again fostering greater employee motivation to change behavior.

The trade-off, however, is that gainsharing plans can pay off big even when the company is losing money. Another difficult situation arises when management would like to bring more work into the plant, but cannot afford to because the plan payouts would become too costly. In these cases, one might say that gainsharing plans (consistent with the general history of incentive plans) sometimes "fail" because they are too "successful." The payouts of any incentive plan must walk the fine line between being too low to motivate employees and being too high for management to afford. Even when standards work well initially, changes in production level and technology often result in the plan being unacceptable to one party or the other. In some cases, management may choose to "buy out" employees by paying a lump sum settlement in exchange for being able to redesign the plan with different standards, especially in unionized settings. An implication is that any sort of variable pay program should have a "sunset" provision that requires evaluation of the plan after a specific number of years, to avoid having the pay program becoming irrelevant because the organization changed, but it did not

A final reason we discuss for the growth in variable pay plans is that the increased use of total quality management (TQM) often entails a movement toward a team-based organization and empowerment of employees to go beyond their traditional roles to make decisions in a broader range of areas that are likely to have an impact on organization performance. Individual-oriented systems may not be adequate for encouraging employees to pursue broad organization goals, and to engage in the cooperative team and group-based decision-making necessary.

A survey conducted for the American Compensation Association (ACA) asked organizations that implemented TQM programs how their pay practices changed (Davis, 1993). As Table 5 indicates, major changes included less reliance on supervisors as the only source of performance appraisals, more reliance on team and organization results in setting pay, greater use of variable pay, and fewer, broader pay grades.

Table 5. Changes in Pay to Support Total Quality

	Before	After
Performance Appraisal (n = 91)		
Only supervisors as source	59 %	12 %
Peer/team appraisals	2 %	25 %
Add quality criteria/goals	--	68 %
Have team goals	--	41 %
Plan Increase Policies (n = 38)		
Increases tied to individual performance appraisals	88 %	60 %
Increases tied to team/organization results	8 %	60 %
Increases tied to quality results	--	49 %
Increases tied to skill/knowledge levels	--	33 %
Incentive Program Policies (n = 56)		
Incentives based on individual results	26 %	31 %
Incentives based on individual/ team results	23 %	37 %
Incentives based on team/organization results	20 %	52 %
Salary Structure Policies		
Other (e.g., more pay at risk)	--	52 %
Fewer grades, broader range widths	--	38 %

Note: n refers to the number of organizations (out of 196 total) that made changes in each pay area.
Source: Davis, J.H. (1993, Autumn). ACA Journal. "Quality Management and Compensation."

From an equity theory perspective, placing the entire employee population on such plans may also create a greater sense of fairness among non-executive employees who typically have not been covered by such plans in the past, but saw that executives were. Of course, this effect may be limited to plans where variable pay is used to provide additional upside earnings potential, as opposed to cases where it replaces a portion of base salary.

Banding, De-Layering, and Paying the Person Rather than the Job

In the traditional pay system, the worth of jobs is assessed on the basis of job evaluation data in combination with market survey data. Job evaluation focuses on measuring and valuing the specific characteristics and requirements of the job. Critics, however, suggest that job-based systems tend to spawn too much bureaucracy, too much emphasis by employees on doing only what is in their job description, and a lack of focus on market comparisons, which are critical for competitiveness. In addition, job levels become status indicators, which can get in the way. For example, an employee may be reluctant to accept a temporary assignment, that would be good from a developmental point of view, unless it has at least as high of a job level.

There have been at least two types of responses. First, organizations like General Electric have cut levels of management and the corresponding pay grades. The goals are to improve communication and speed decisions by reducing the levels of management, and to provide wider pay grades (or bands) in order to allow more flexibility to recognize individual contributions, and to make lateral movements simpler by reducing the likelihood of a job being in a different (in this case, lower) grade (and looking like a demotion).

The participants in the Dyer and Blancero (1993) study were also asked how the number of pay levels in the hypothetical service organization would change by the year 2000. Across the four occupational groups, the 36 pay levels in 1991 were expected to decrease to 23 pay levels by 2000, a decrease of about one-fourth. Whether the hoped for advantages of delayering and banding will offset the potential drawbacks (e.g., less opportunity for promotion) remains to be seen.

Aside from allowing more flexibility in moving employees, banding, by virtue of a greater spread between the minimum and maximum in each pay grade, is also intended to provide more opportunity to recognize individual differences in performance. So, within-level pay growth for high performers will increase, while promotion opportunities and related pay growth will decrease. It remains to be seen whether this will, on balance, be a good trade for motivational purposes. Further, banding carries the risk of becoming very expensive. Topping out of employees near the maximum would be very expensive under a banding system. Some organizations have implemented sub-bands or zones within bands to avoid this problem. However, one might then reasonably ask what the difference is between an old system with 30 grades and a new system with 10 bands, each with 3 sub-bands

Another trend is for some organizations to move away from linking pay to job content through job evaluation, and instead pay workers for the skills they possess. Skill-based pay links pay to the breadth or depth of employee skill. The goal is to encourage learning, which in turn facilitates flexibility in work assignments and encourages learning as a way of life to help with future organization change.

Where are these recent developments in pay likely to lead us? We know that money can be a powerful motivator. Indeed, a literature review of four motivational programs (individual monetary incentives, goal-setting, job redesign, and participation in decision making) found that monetary incentives were associated with the largest average increase in physical productivity (Locke, Feren, McCaleb, Shaw, & Denny 1980). Therefore, changes in pay practices have the potential to significantly change attitudes, behaviors, and organization functioning. The challenge, however, is to realize the potential of money as a motivator without running afoul of the many roadblocks that arise in terms of measuring performance, setting standards that are perceived as fair, and choosing the right mix of individual, group, and organization objectives to reward.

As one recent example of a variable pay program gone wrong, consider the problems Sears encountered in some of its automotive repair shops in New Jersey and California. In a State of California undercover investigation, 38 visits to 27 Sears repair shops resulted in 34 cases of unnecessary service or repair recommendations. Edward A. Brennan, the chairman of Sears, stated that "the incentive compensation program and sales goals created an environment where mistakes occurred" (Fisher, 1992). In essence, repair shop employees had been rewarded for driving revenue (i.e., selling repairs to customers). Sears subsequently changed its pay system to one that focused on "quality."

Although specific examples are useful to demonstrate specific points, what does the broader research literature tell us regarding the typical outcomes of variable pay and other pay for performance programs?

At the organization level, evidence suggests that greater emphasis on short-term bonuses and long-term incentives (relative to base pay) is associated with higher subsequent profitability, at least among top and middle level managers (Gerhart & Milkovich, 1990). Specifically, an organization with a bonus/base ratio of 10%, and 28% of its managers eligible for long term incentives had an average return on assets of 5.2%. In contrast, an organization with a 20% bonus to base ratio, and 48% of its managers eligible for long term incentives, had an average return on assets of 7.1%.

The fact that organization-based bonuses and incentives work for high-level managers does not necessarily mean they will work for other types of employees, most of whom have less influence over organization performance and thus, weaker instrumentality perceptions. Still, even if the motivational impact (in terms of sheer effort) of organization-based incentives is weaker for such groups, cost control and a re-focusing of behavior toward broader organizational goals may still be possible with such programs.

The empirical evidence on profit sharing plans, in fact, generally paints a positive picture, with organizations using profit sharing having higher productivity (usually defined as value added per employee) on average than organizations that do not use profit sharing (e.g., Weitzman & Kruse, 1990; Kruse, 1993a, 1993b). Still, there has yet to be a convincing demonstration that profit sharing actually causes better organization performance (Gerhart & Milkovich, 1992). It may be that organizations with higher profit levels are more likely to adopt profit sharing plans. In addition, if a profit sharing plan does not work out, it is likely to be discontinued. So, the only profit sharing plans that are studied are those that have proven successful, and we do not hear about the plans that failed or needed to be replaced after they served their purpose. An organization that is deciding whether to adopt a profit sharing plan must know how often such plans work and how often they fail or get discontinued, not just how well the successful plans work.

The evidence on stock plans is very limited, aside from the Gerhart and Milkovich (1990) study of top and middle level managers. The evidence that is available pertains mostly to employee stock ownership plans (ESOPs). Like profit sharing, the evidence is generally favorable (Jones & Takao, 1993; Conte & Svejnar, 1990), but the same cautions regarding causality apply. In any case, research suggests any beneficial effects of ESOPs may be stronger where employees have greater participation in making decisions, perhaps because it gives the employee a stronger feeling of ownership (Pierce, Rubinfeld, & Morgan, 1991). The costs of stock plans, especially options, may not always be obvious, but purchasing stock or issuing new stock (and the resulting dilution of the value of other shares) are costly moves. Indeed, U.S. Senator Carl Levin of Michigan introduced a bill in January 1993 that would require companies to show the granting of stock options as an expense. The Financial Accounting Standards Board (FASB) also has proposed changes to its rules in this area.

What is the evidence on gainsharing programs? Again, it is generally positive. Although the types of cautions cited above regarding causality apply, the fact that from a theoretical standpoint, gainsharing programs offer employees a better line of sight (or instrumentality) between their performance and rewards (Lawler, 1989; Schwab, 1973) suggests that the motivational impact of such programs may be stronger than is the case with organization-wide programs, like profit sharing and stock plans. Gainsharing payouts are typically based on measures like value added, sales value of production, or hours saved, which are more controllable by employees than profits or stock performance.

A time series study by Schuster (1983) of six gainsharing plans found substantial (around 30 %-) increases in productivity in four cases following the implementation of gainsharing. A fourth plant had an initial increase in productivity, but increases in the costs of raw materials subsequently decreased the value added per worker, leading to no bonuses. A fifth plant, although not showing an increase in productivity, had gainsharing in place for twenty years, suggesting that the plan was working, but the productivity increase had already occurred before the study. Other studies have also found significant productivity improvements from gainsharing programs (e.g., Kaufman, 1992).

In addition to having a payout measure that is controllable, gainsharing plans often have the advantage of covering a smaller number of employees, which is also beneficial for motivation, because there is less likelihood of employees "free riding" (i.e., working less hard because others will work hard). Indeed, one study estimated that a doubling of employees covered by a gainsharing plan from around 200 to 400 would reduce the expected productivity gain by almost one-half (Kaufman, 1992). The implication is that the number of employees covered can have a substantial impact on the plan's success.

The fact that gainsharing (or any pay program) has a positive impact on productivity is no guarantee that it will continue to be used. A study of a gainsharing plan at an electrical utility estimated a net savings of between $857,000 and $2 million, but the plan was discontinued because employees in other divisions (all represented by the same union) felt unfairly treated because they were not covered (Petty, Singleton, & Connell, 1992). The organization was then faced with two difficult options. First, it could include all employees under the same plan, but that would likely increase the free rider problem and reduce the motivational impact. Second, it could have a separate plan for each division, but this could easily result in unequal payoffs to employees in different divisions, raising the same problems originally encountered with employees and the union. There would also need to be a means of preventing between-division competition. A profit sharing or stock plan combined with gainsharing plans would be one option

Other evidence also indicates that plans which appear to save money do not necessarily survive very long. Kaufman (1992) found that discontinued plans had improved productivity nearly as much as continuing plans. A study sponsored by the American Compensation Association (McAdams & Hawk, 1992) may shed some light on this question. They found that gainsharing plans, on average, were associated with net gains per employee of between $1,300 and $3,700 per year. Nevertheless, when asked to rate the effectiveness of gainsharing plans in improving effectiveness in areas like business performance, fostering teamwork, strengthening the pay-for-performance link, and so forth, the average effectiveness ratings all fell between 2.63 and 3.25 on a 1 (no effectiveness) to 5 (high effectiveness) scale. In other words, most respondents were pretty lukewarm about gainsharing.

Many organizations are moving to group and organization variable pay plans because they are frustrated with what they see as the failure of more traditional merit pay plans. Commonly cited problems include a lack of adequate differentiation between good and poor performers, employee and supervisor resistance, and the fact that merit increases sometimes seem to have become viewed as an entitlement by employees that is costly, and does not vary with business performance.

Although there is truth to many of these assertions, one sometimes wonders if perhaps merit pay has been pronounced dead too soon. So-called studies of merit pay have often had significant limitations (see Gerhart & Milkovich, 1992). In addition, the notion that there is no merit pay is open to question.

It is common to conclude that there is no individual merit pay because raises received by good and poor performers differ by only a few percentage points. Two employees, each with a base salary of $40,000, one receiving a 5% increase, the other 6%, would receive raises differing by $400 per year before taxes, or about $8 per week. Framed this way, the difference does indeed seem small and unlikely to motivate performance.

On the other hand, the example ignores the fact that high performers are more likely to be promoted and thus, will have greater earnings growth. This is part of pay-for-performance, but it may not always be communicated as well as it could. Further, even limiting one's attention to the annual increase process, it can be shown that small differences in pay raises accumulate into significant differences over time. As Table 6 shows, the present value (or "real" payoff) to raises higher by 1 percentage point adds up to about $76,000 over 20 years. Factoring in promotion based on performance and pay-linked benefits (e.g., retirement) would further increase the payoff to higher performers. Factoring in taxes would decrease the payoff. Would communicating the payoff to performance in this manner change the way employees react to merit pay? Our conversations with managers yields a wide array of opinions on the matter, suggesting a good area for future research.

		Employee 1		**Employee 2**	
Performance Rating		Average		1 point above above average	
Annual Pay Growth		5%		6%	
Year		Nominal	Real*	Nominal	Real**
	1	$40,000		$40,000	
	2	$42,000	$40,000	$42,400	$40,381
	3	$44,100	$40,000	$44,944	$40,766
	4	$46,305	$40,000	$47,641	$41,154
	5	$48,620	$40,000	$50,499	$41,546
	6	$51,051	$40,000	$53,529	$41,941
	7	$53,604	$40,000	$56,741	$42,341
	8	$56,284	$40,000	$60,145	$42,744
	9	$59,098	$40,000	$63,754	$43,151
	10	$62,053	$40,000	$67,579	$43,562
	11	$65,156	$40,000	$71,634	$43,977
	12	$68,414	$40,000	$75,932	$44,396
	13	$71,834	$40,000	$80,488	$44,819
	14	$75,426	$40,000	$85,317	$45,245
	15	$79,197	$40,000	$90,436	$45,676

	16	$83,157	$40,000	$95,862	$46,111
	17	$87,315	$40,000	$101,614	$46,551
	18	$91,681	$40,000	$107,711	$46,994
	19	$96,265	$40,000	$114,174	$47,441
	20	$101,078	$40,000	$121,024	$47,893
Total		$1,322,638	$760,000	$1,471,424	$ 836,690
Difference				$ 148,785	$ 76,690

*Using 5 % discount rate.

The empirical evidence on the effects of banding is basically non-existent. Research on skill-based pay is just beginning to emerge. An ACA survey of organizations using skill-based pay illustrates some of the potential advantages and disadvantages. Most survey respondents felt that skill-based pay was successful in contributing to greater workforce flexibility and adaptability and in supporting work teams. However, relatively few saw any reduction in labor costs or layoffs. (Indeed, skill based pay is thought to permit a leaner headcount because of cross-training.) So, one must consider whether possible higher labor costs are justified by the advantages having to do with flexibility, adaptability, and the use of teams. Further, it must be recognized that if a plan is implemented, there are several factors that can contribute to its termination. The ACA survey found the following to be most important in terminating skill based pay plans: inadequate management commitment, unwillingness to endure short-term implementation problems, poor plan designs that increase labor costs without providing offsetting organizational benefits, conflicts between employees included and those excluded from the plan, inadequate training opportunities, and the failure of management to require meaningful skill certifications prior to pay increases.

Only one skill based pay study to date (Murray & Gerhart, 1994) has used objective measures of productivity and quality, a control group, and a time series before and after implementation of the plan. In a comparison of two automobile parts plants, Murray and Gerhart found that a significant increase in productivity and product quality took place in the plant that implemented skill based pay.

GLOBALIZATION AND COMPENSATION

The continued globalization of markets means that we will have to increasingly consider whether the effect of different pay strategies is likely to differ from country to country, or between cultures within a country.

Table 7. National Culture Clusters

Region or Country	Power Distance	Individualism	Masculinity	Uncertainty Avoidance
Pacific Rim				
Hong Kong, Malaysia, Philippines, Singapore	High	Low	High	Low
Japan	High	Low	High	High
South Korea, Taiwan	High	Low	Low	High
United States, Great Britain	Low	High	High	Low

Sources:Hofstede, G. (1993, February). Cultural constraints in management theories. Academy of Management Executive, 7(1), 81-94; Hodgetts, R.M. & Luthans, F. (1993, March-April). U.S. multinationals' compensation strategies for local management: Cross-cultural implications. Compensation & Benefits Review, 25, 42-48. Review, 25, 42-48.

Hofstede's (1993) work on identifying culture differences on dimensions such as power distance (i.e., the degree of inequality considered normal), individualism, masculinity, uncertainty avoidance; and short- versus long-term time orientation has been used by Hodgetts and Luthans (1993) to begin studying this question. Certain hypotheses flow readily from the national differences depicted in Table 7.4 Variable pay (pay at risk) may face difficulties in countries that have a high need for uncertainty avoidance such as Japan, South Korea, and Taiwan. Individualistic programs such as merit pay could be a problem in cultures where collectivism is a stronger norm than individualism (e.g., the Pacific Rim countries). Still, average differences in culture are just that, averages, and should not necessarily be viewed as factors that must be taken as a given. Honda, in Japan, for example, just recently announced that it would be changing many of its managers over to a merit pay system.

On the other hand, U.S. companies that have attempted to export pay practices overseas have often encountered difficulties. Lincoln Electric, famous for its history of success using variable pay, has thus far not been successful in implementing variable pay in its overseas acquisitions. Our own experience with gainsharing plans in Western Europe has not been successful (Chilton, 1993). The cultural differences described by Hofstede and related customs are often difficult to overcome. It is probably significantly easier to implement pay practices that are not typical of a country in a greenfield setting as opposed to an acquisition. Indeed, Japanese (e.g., Honda, Nissan) and German (e.g., BMW,Mercedes-Benz) automobile plants opened in the United States have often been in greenfield sites, where the company has maximum flexibility in screening and choosing employees who will fit well with their corporate culture, human resource management, and pay philosophies.

CONCLUSION

Our goal in this chapter has been to describe the theory and practice of compensation, as well as provide an overview of recent empirical evidence on the consequences of different compensation practices. The theory section points to the many trade-offs in designing employee compensation policies. Examples of trade-offs include maximizing high individual effort versus teamwork and cooperation, controlling costs versus maximizing employee effort, and providing incentives for promotion versus producing feelings of inequity due to large pay differentials. Our message has been that the nature of such trade-offs should depend on the corporate and business strategies and that the trade-offs can be made less of a problem by combining pay programs in a way that helps balance competing objectives.

As a final comment, we would like to emphasize that, although it is important to keep abreast of what other organizations are doing (benchmarking) in the area of employee compensation, it is crucial to remember that what works for one organization may not work at all for another. Therefore, surveys of "best practices" are useful to the extent that the surveys report a diversity of best practices and the reasons why different practices are best for different organizations. The ultimate choice of a best compensation strategy rests, of course, on its fit with other human resource activities and its fit with the business strategy

REFERENCES

1. Adams, J.S. (1963). Toward an understanding of inequity. Journal of Abnormal Psychology, 67, 422-436.
2. Beam, B.T. Jr., & McFadden, J.J. (1992). Employee benefits. Chicago: Dearborn Financial Publishing.
3. Bretz, R.D., Ash, R.A., & Dreher, G.F. (1989). Do people make the place? An examination of the attraction-selection-attrition hypothesis. Personnel Psychology, 42, 561-581.

4. Chilton, K. (1993, November-December). Lincoln Electric's incentive system: Can it be transferred overseas? Compensation and Benefits Review, 21-30.

5. Conte, M.A., & Svejnar, J. (1990). The performance effects of employee ownership plans. In A.S. Blinder (Ed.), Paying for productivity, pp. 245294. Washington, D.C.: The Brookings Institution.

6. Cowherd, D.M., & Levine, D.I. (1992). Product quality and pay equity between lower-level employees and top management: An investigation of distributive justice theory. Administrative Science Quarterly, 37, 302-320.

7. Davis, J.H. (1993, Autumn). ACA Journal. Quality management and compensation.

8. Dyer, L., & Blancero, D. (1993). Workplace 2000: A delphi study. Center for Advanced Human Resource Studies, Cornell University.

9. Eisenhardt, K.M. (1989). Agency theory: An assessment and review. Academy of Management Review, 14, 57-74.

10. Ellig, B.R. (1981). Compensation elements: Market phase determines the mix. Compensation Review, (Third Quarter), 30-38.

11. Fama, E.F., & Jensen, M.C. (1983). Separation of ownership and control. Journal of Law and Economics, 26, 301-325.

12. Fisher, L.M. (1992, June 23). Sears auto centers to halt commissions. New York Times, D1.

13. Gerhart, B., & Milkovich, G.T. (1990). Organizational differences in managerial compensation and financial performance. Academy of-Management Journal, 33, 663-691.

14. Gerhart, B., & Milkovich, G.T. (1992). Employee compensation: Research and practice. In M.D. Dunnette , & L.M. Hough (Eds.), Handbook of industrial and organizational Psychology, 2nd Edition, pp. 481-569. Palo Alto, CA: Consulting Psychologists Press, Inc.

15. Gerhart, B., Milkovich, G.T., & Murray, B. (1992). Pay, performance, and participation. In D. Lewin, O. Mitchell, & P. Sherer (Eds.), Research Frontiers in Industrial Relations, pp. 193-238. Madison, WI: Industrial Relations Research Association.

16. Gomez-Mejia, L.R. (1992). Structure and process of diversification, compensation strategy, and firm performance. Strategic Management Journal, 13, 381-397.

17. Gomez-Mejia, L.R., & Balkin, D.B. (1992). Compensation organizational strategy and firm performance. Cincinnati: South-Western

18. Greenberg, J. (1990). Employee theft as a reaction to underpayment of inequity: The hidden cost of pay cuts. Journal of Applied Psychology, 75, 561-568.

19. Heneman, H.G., III , & Schwab, D.P. (1979). Work and rewards theory. In D. Yoder , & H.G. Heneman, Jr. (Eds.), ASPA handbook of personnel and industrial relations. Washington, D.C.: Bureau of National Affairs.

20. Hodgetts, R.M., & Luthans, F. (1993, March-April). U.S. multinationals' compensation strategies for local management: Cross-cultural implications. Compensation & Benefits Review, 25, 42-48.

21. Hofstede, G. (1993, February). Cultural constraints in management theories. Academy of Management Executive, 7(1), 81-94.

22. Jones, D.C., & Takao, K. (1993). The scope, nature, and effects of employee stock ownership plans in Japan. Industrial and Labor Relations Review, 46, 352-367.

23. Judge, T.A., & Bretz, R.D. Jr. (1992). Effect of values on job choice decisions. Journal of Applied Psychology, 77, 261-271.

24. Kaufman, R.T. (1992). The effects of Improshare on productivity.- Industrial and Labor Relations Review, 45, 311-322.

25. Kerr, J.L. (1985). Diversification strategies and managerial rewards. Academy of Management Journal, 28, 155-179.

26. Kidwell, R.E. Jr., & Bennett, N. (1993). Employee propensity to withhold effort: A conceptual model to intersect three avenues of research. Academy of Management Review, 18, 429-456.

27. Kruse, D.L. (1993a). Profit sharing: Does it make a difference? Kalamazoo, MI: Upjohn Institute.

28. Kruse, D.L. (1993b). Does profit sharing affect productivity? Tests using panel data on profit sharing features and personnel policies. Working Paper, Rutgers University.

29. Lambert, R.A., & Larcker, D.F. (1989). Executive compensation, corporate decision-making, and shareholder wealth. In F. Foulkes (Ed.), Executive compensation (pp. 287-309). Boston: Harvard Business School Press.

30. Lawler, E.E. III. (1989). Pay for performance: A strategic analysis. In L.R. Gomez-Mejia (Ed.), Compensation and benefits. Washington, D.C.: Bureau of National Affairs.

31. Levine, David I., & Tyson, Laura D. (1990). Participation, productivity, and the firm's environment. In Alan S. Blinder (Ed.), Paving for productivity. Washington, D.C.: Brookings Institution.

32. Locke, E.A., Feren, D.B., McCaleb, V.M., Shaw, K.N., & Denny, A.T. (1980). The relative effectiveness of four methods of motivating employee performance. In K.D. Duncan, M.M. Gruenberg, & D. Wallis (Eds.), Changes in working life (pp. 363-388). New York: Wiley.

33. McAdams, J.L., & Hawk, E.J. (1992). Capitalizing on human assets through performance-based rewards. ACA Journal, 1(1), 60-73.

34. Miles, R.E., & Snow, C.C. (1978). Organizational strategy, structure, and process. New York: McGraw-Hill.

35. Milkovich, G. T. (1988). A strategic perspective on compensation management. Research in Personnel and Human Resources Management, 6, 263-288.

36. Milkovich, G.T., Gerhart, B., & Hannon, J. (1991). The effects of research and development intensity on managerial compensation in large organizations. Journal of High Technology Management Research, 2, 133-150.

37. Milkovich, G.T., & Newman, J. (1993). Compensation (4th Ed.). BPI/Irwin: Homewood, IL

38. Milkovich, G.T., & Wigdor, A.K. (1991). Pay for performance. Washington, D.C.: National Academy Press.

39. Murray, B., & Gerhart, B. (1993). Organizational outcomes from the introduction of a skill-based pay program. Working Paper. Center for Advanced Human Resource Studies, Cornell University.

40. Noe, R.A., Hollenbeck, J.R., Gerhart, B., & Wright, P.M. (1994). Human resource management: Gaining a competitive advantage. Burr Ridge, IL: Austen Press/Irwin.

41. Petty, M.M., Singleton, B., & Connell, D.W. (1992). An experimental evaluation of an organizational incentive plan in the electric utility industry. Journal of Applied Psychology, 77, 427-436.

42. Pierce, J.L., Rubenfeld, S., & Morgan, S. (1991). Employee ownership: A conceptual model of process and effects. Academy of Management Review, 16, 121-144.

43. Santora, J.E. (1991, February). DuPont returns to the drawing board. Personnel Journal, 34-36

44. Schuster, M. (1983). The impact of union-management cooperation on productivity and employment. Industrial and Labor Relations Review, 36, 415-430.

45. Schwab, D.P. (1973). Impact of alternative compensation systems on pay valence and instrumentality perceptions. Journal of Applied Psychology, 58, 308-312.

46. Tosi, H.L. Jr., & Gomez-Mejia, L.R. (1989). The decoupling of CEO pay and performance: An agency theory perspective. Administrative Science Quarterly, 34, 169-189.

47. Tully, S. (1993, November 1). "Your Paycheck Gets Exciting." Fortune, p. 83+.

48. Vroom, V.H. (1964). Work and motivation. New York: Wiley.

49. Weitzman, M.L., & Kruse, D.L. (1990). Profit sharing and productivity. In A.S. Blinder (Ed.), Paying for productivity. Washington, D.C.: Brookings Institution.

Chapter 13

Role of Job Satisfaction in the relationship of Business Excellence and OCB: Iranian Hospitality Industry

Huseyin Araslı, Sarvnaz Baradarani

**Eastern Mediterranean University School of Tourism & Hospitality Management Gazimagusa, North Cyprus Via Mersin 10, Turkey*

ABSTRACT

The main aim of the study was to evaluate the relationship between leadership, management by fact, continuous improvement, all work is process, people based management, teamwork with job satisfaction of the frontline employees of 3, 4 and 5 star hotels in Tehran, as well as the relationship of the job satisfaction with the organizational citizenship behavior. The study contained 314 questionnaires which were collected from the frontline employees of the hotels in Tehran. The data were analyzed using PLS in order to evaluate the relationships between the business excellence practices with job satisfaction and organizational citizenship behavior. Having adopted the questionnaires it has been investigated that only leadership and teamwork had non-significant effect on job satisfaction of the frontline employees working in three, four and five star hotels of Tehran that led to organizational citizenship behavior. The study suggests various implications for the hotel managers, politicians and policy makers of Iran, in order to achieve higher levels of job satisfaction as well as organizational citizenship behaviors in their hotels.

Keywords: Business Excellence Practices, Job Satisfaction, Organizational Citizenship Behavior, Iran, Hotels.

1. INTRODUCTION

Organizations need to implement a proper quality management system in order to deal with the changes in today's fast growing industries. The tourism industry is known to be one of the fastest growing industries in the new millennium. Iran with its rich history is one of the gradually developing countries in terms of tourism. The main resource for revenue gain is the petroleum products of Iran; however tourism is also one of the earning industries for such a country. Iran is ranked 7th in the world in terms of its historical places and monuments, and it is known to be 70th in term of its tourism hosting capacity (WTTC, 2010). In terms of world tourism revenues, Iran's share is a meager 0.000%. In 2004, a total number of 1,411,500 tourists visited Iran, 431,500 of whom were foreigners and the rest were Iranian citizens (ICHTO, 2006).

The secretary of the Iran Hotel Owners Society (Iran Daily, 2006) of Iran declared that, there are 619 hotels in the country, with a total capacity of 27,034 rooms and 55,382 beds. These hotels include 17 five-star hotels, 27 four-star hotels, 60 three-star hotels, 203 two-star hotels, 212 one-star hotels, and 100 not graded hotels (Ministry of Culture and Islamic Guidance, 2000). These establishments have generated US$29.8 billion of economic activity in 2006, and it was estimated to grow toUS$71.4 billion by 2016. Total demand is expected to grow by 2.4% and by 4.0% per annum through 2016.

A rapid growth in the country's hotel industry is planned, so there are new hotels being constructed and old hotels are being renovated in order to be able to compete with other hotels in the hospitality sector. Iran's travel and tourism industry are expected to contribute 3.5% to the Gross Domestic Product (GDP) (US$8.4 billion), rising in nominal terms to US $23.6 billion (4.0% of total) by 2016. The travel and tourism economy involvement should decline from 9.8% -US $23.4 billion- to 9.1% -US $53.6 billion- during this same period.

In terms of employment there are 1,712,000 jobs in the Iranian travel and tourism industry which contains 8.7% of the employment share of Iran. The industry is expected to generate 12.1% of total exports US $8.9 billion, growing to US $13.2 billion (19.4% of total) by 2016 (WTTC, 2006). Based on the aforesaid statistical results, the administrations in the industry should use necessary management tools in order to get better outcomes and avoid low levels of job satisfaction and organizational citizenship behaviors.

One of the most important contributions of this study to the hospitality literature is to mention the importance of business excellence practices and their relationships with job satisfaction and organizational citizenship behavior in the hospitality industry, since there are very limited studies that have taken the business excellence practices into consideration in the hotels, especially in a developing country like Iran (Oakland and Tanner, 2008; Arasli and Baradarani, 2010)

2. LITERATURE REVIEW

The fast growing changes in the organizations have resulted in the need for quality management systems which led to developing various quality models, since different awards were devoted to the best models many models developed. Deming Prize was one of the first models that was conducted in Japan (1951), which was developed by the board of Directors of the Japanese Union of Science and Engineering. Whereas the European Foundation for Quality Management (EFQM) model was developed in the western world in 1991 and the USA's Malcolm Baldrige Award (MBNQA) introduced in 1987 were the other famous models.

In addition there were different definitions for total quality management as well. For example Drucker (1964) defined total quality management as an organizational performance which means how well organizations do their job. Dale et al, (1994) defined it as "to extend the scope of the Business Excellence theory it is necessary to incorporate management theories into its development and that much remains to be done for Business Excellence to reach a stage of 'refine/extend' in the theory building process" (Dale et al, 1994).

Arasli (2002) is one other scholar who stated an expression for TQM; he defines total quality management as "the satisfaction of social-shareholders; customers, staff, owner/s and suppliers by implementing hard issues, such as effective planning, programmers, policies and strategies, as well as using soft issues, such as human and all other assets, efficiently and continually within an organization" (Arasli, 2002).

Enrique Claver-Cortes et al. (2007) mentioned that TQM and business excellence are management systems that influences the organizational performances in both manufacturing and service organizations. As the researches indicate; the numbers of business excellence studies made in the tourism sector are very few, while the numbers of business excellence studies in the hospitality industry are even more limited. Koc (2006) and Hudson's (2004) are two scholars who studied business excellence in the tourism industry, they evaluated the tourism product as a package of several products, including transportation, accommodation, food and various activities and the tourist consumption of a total experience, in which total quality management has been identified as the key subject in differentiating service products and building competitive advantage in tourism

There are some other studies who have concentrated on the business excellence dimension in the hospitality industry which are conducted in Spain and Cyprus (Camison, 1996; Soriano, 1999; Arasli, 2002) and also in benchmarking national tourism Organizations and agencies (Holmes, 2007).

Kanji (2002) is also one other scholar who has a well-known model for the business excellence and he defined business excellence as a mean that measures the satisfaction level of the customers, employees and the stakeholders at the same time in the organization in order to obtain a comprehensive evaluation of the performance of the organization (Kanji, 2002). This study has taken some dimensions of Kanji's business excellence practices as well as the dimensions used in the business excellence practices of Arasli (2012).

Kanji has developed a practical business excellence model for the organizations; the model can be adapted to the organization's needs, with high flexibility. Kanji's easily adaptable model is an extremely cost and time efficient. It applies comprehensive statistical analysis and it has a strong focus on improvements rather than positioning the organization's quality management scheme (Kanji, 2002).

Leadership, management by fact, all work is process, teamwork, people based management and continuous improvement are the business excellence practices that are evaluated in this study by considering their relationships with the job satisfaction and organizational citizenship behaviors of the frontline employees in the hotels of Tehran.

LEADERSHIP

Leadership is a very important criterion in the quality models, especially in the business excellence model and Malcolm Baldrige model (Wong, 2001) and is defined as a quality culture that relates to the vision, mission, and goals of the organization. Leadership aims to establish shared values and launch systems to increase customer satisfaction. It also communicates, defines, and motivates continuous improvement (Kanji, 2002). Leadership skills are very essential for the hospitality industry considering its future competition, dynamic environment, service orientation, and labor-intensive nature (Gillet & Morda, 2003; Pine & Wong, 1998).

Management by Fact

Leadership is a very important criterion in the quality models, especially in the business excellence model and Malcolm Baldrige model (Wong, 2001) and is defined as a quality culture that relates to the vision, mission, and goals of the organization. Leadership aims to establish shared values and launch systems to increase customer satisfaction. It also communicates, defines, and motivates continuous improvement (Kanji, 2002). Leadership skills are very essential for the hospitality industry considering its future competition, dynamic environment, service orientation, and labor-intensive nature (Gillet & Morda, 2003; Pine & Wong, 1998).

Management by Fact

Management by fact evaluates quality improvement procedures using the organization's performance measurement system. In addition, it distributes performance measurement throughout the whole organization in order to improve its products and processes (Kanji, 2002). Once discussed, the management should use the best knowledge available to make decisions as described by Juran et al. (1988).

All Work is Process

The organization must follow a process-oriented method that focuses on how the organization is working rather than the results of the work. The quality requirements should be assessed; this assessment should be used to enhance knowledge about the processes. Statistical methods are applied in order to control the processes (Kanji, 2002). However in order to be more effective, the quality control techniques are essential in having a better business excellence should not only be considered to be used by the specialists, but they should rather be used in everyday life of the employees (Kanji, 2008).

Teamwork

People in the organization should be provided with the opportunity to work together as a team and should be aware of the importance of co-dependent tasks and enhanced communication (Kanji, 2002). The effective teamwork in an organization leads to a common goal that can increase the motivational level of the employees as well as job satisfaction (Griffin, Patterson, & West, 2001). Teamwork is considered to be the source of employee autonomy, meaningfulness, bonding with team members, and satisfaction (Denison & Hart, 1996; Mitchell et al., 2001).

People-Based Management

In people-based management, employees should feel that they play an important role in the organization's success. Employee's contributions are valued by the organization and feedback is provided on their performance, which is directly reflected in the quality of the product or service provided by the organization (Kanji, 2002). The employees should be provided with the necessary resources as well as rewards when needed. The employees perform better if they work in a more friendly and cooperative environment (Kanji, 2008).

Continuous Improvement

Continuous improvement observes changes in customer orders. The organizations look for ways to improve. Quality improvement methods are used to improve all products, services, and processes; the organizations have to compare its current quality levels of service with competitors. Continuous improvement dimension is a conductive process to learn, and it requires an organizational culture in which all the employees constantly become innovative. Improvement is the result of learning from the previous mistakes and implementing corrective actions as well as trying new things from the feedback they have gained (Kanji, 2008).

Job Satisfaction

Job satisfaction has been defined by many scholars, and the main definition is the differences that exist between the expectations about the job, and the actual attribute of the job (Heslop et al, 2002). Locke (1969, p. 317) is one of the scholars who defined job satisfaction by mentioning that; job satisfaction is the pleasant feelings that results from the appraisal of the job or by the job facilities, whereas job dissatisfaction is the unpleasant feelings that results from the appraisal of job such as, frustration or blocking the achievement of the values (Schwepker, 2001, p. 41). Herzberg et al. (1959) was one other scholar who had submitted the main theory of job satisfaction (Herzberg et al., 1959). He divided the needs of the employees to two which were; hygiene and motivation. Hygiene factors satisfy the employees in certain conditions such as supervision, interpersonal relations, physical working conditions, salary, benefits, etc. However it has been mentioned that these factors not fully satisfy the employees, and they only reduce the dissatisfaction level (Furnham et al., 2002). Job satisfaction is formed by the intrinsic and extrinsic satisfaction factors. The intrinsic factors are related to the ability utilization, activity, achievement, authority, independence, moral values, responsibility, security, creativity, social service, social status, and variety. Whereas the extrinsic job satisfaction factors are advancement, company policy, compensation, recognition, supervision-human relations, and supervision-technical (Feinstein & Vondrasek, 2001). Organizational Citizenship Behaviour

Organizational citizenship behaviour (OCB) has become very popular in psychology and management, and it has been concentrated by many scholars (Bateman et al, 1983; Organ, 1988; Smith et al., 1983; Podsakoff et al., 2000; Tang et al., 2008). OCB refers to the flexible individual behavior that is not directly recognized by the formal reward system and combines the efficient and effective functioning of the organization (Organ, 1988). In the hotel environment, service-oriented OCB promotes a more effective service delivery procedure, provides better service quality, establishes a more friendly customer interaction, and offers more innovative thoughts about services for better serving the customers and scoring a greater customer satisfaction (Podsakoff et al., 2009; Raub, 2008; Stamper and Van Dyne, 2001; Yen and Niehoff, 2004). Therefore, hotels have started to seek useful means to enhance service-oriented behaviors of their customer contact employees, because this type of behaviors is beneficial for service delivery quality, competitive advantages, and financial performance of the hotels (Hartline & Ferrell, 1996; Karatepe et al., 2006; Walz & Niehoff, 2000). However, up till now only few studies have investigated antecedents that promote service-oriented OCB in the hotel industry (Kim et al., 2010; Ma & Qu, 2011; Tsaur & Lin, 2004).

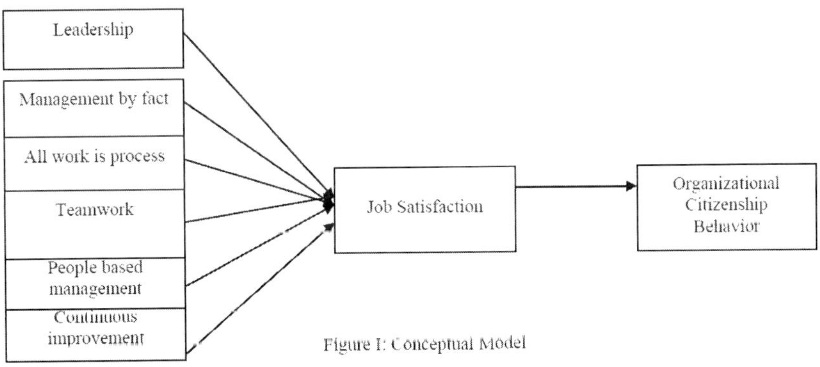

Figure I: Conceptual Model

3.RESEARCH METHOD

3.1.Sample

The front office employees of three-, four- and five star hotels of Tehran - the capital city of Iran - are the main research environment that has been evaluated in the study. The total numbers of 3, 4 and 5 star hotels that are located in Tehran are 32, of which 6 of them are five-star hotels, 11 four-star and 15 three-star. From the total number of 32 hotels only 25 hotels have been used in this study. The remaining 7 hotels did not accept the appointments made by the research team. The data were collected through questionnaires that were conducted through self-administered questionnaires by the research team. Among the 350 questionnaires distributed, 314 were collected individually in 3 weeks in January 2013.

3.2.Measurement

The business excellence quality culture improvement survey is a self-administered perception survey, a self-administered questionnaire addressed to the frontline employees of targeted hotels. The questionnaires were first prepared in English and then translated into Farsi, using back to back translation method. Respondents were asked to indicate their degree of agreement with each system on a five point Likert scale: 1 = strongly disagree, 2 = agree, 3 = somewhat agree / disagree, 4 = agree, and 5 = strongly agree.

All of the variables were measured as it has been covered in table II, that are captured indirectly with direct measurement items. Leadership was measured with the scales developed by Arasli and Baradarani (2010). Continuous Improvement, Management by Fact and All work is Process and Teamwork was measured by Kanji (2002). Job Satisfaction was measured by Hartline and Ferrell (1996). Organizational Citizenship Behavior was measured by Williams and Anderson (1991).

Table I: Demographic Characteristics of the Sample (n=314)

Demographic Variable	Sample Composition	
Gender	Men	63.4%
	Women	36.6%
Hotel Star	3- Star	33.1%
	4- Star	44.9%
	5- Star	22%
Education	High School Diploma	16.6%
	Two Year Diploma	32.8%
	Bachelors	40.4%
	Masters	10.2%
Work Experience	Less than 1 year	10.2%
	1-3	22.6%
	4-6	28%
	7-10	23.9%
	More than 10 years	15.3%

Table II. Measurement

Constructs	Items	References
Leadership	5	Arasli, H. and Baradarani, S. (2011)
Continuous Improvement	5	Kanji, G. K. (2002)
Job Satisfaction	5	Hartline and Ferrell (1996)
Management by Fact	3	Kanji, G. K. (2002)
OCB	5	Williams and Anderson (1991)
People Based Management	3	Kanji, G. K. (2002)
All Work is Process	6	Kanji, G. K. (2002)
Teamwork	3	Kanji, G. K. (2002)

4 DATA ANALYSIS AND RESULTS

A structural equation modeling was used by Partial Least Squares (PLS) approach in order to analyze the relations between the constructs. The samples used in this study were 314, and the Smart PLS (Hansmann & Ringle, 2004) was used for the analysis. There are two steps that have to be conducted in the analysis done by PLS. In the first step validity and reliability tests are submitted in order to test the measurement models. In the second step the structural model is examined by the paths between the constructs of the model. The second step results in the predictions made about the model to ensure the reliability and validity of the model before the conclusions have been made about the different constructs (Hulland, 1999).

4.1.Reliability and Convergent Validity

The reliability of the data was examined through composite reliability values, as it has been indicated in Table III, all the cronbach's alphas are between 0.94 and 0.86, which are all above the cut-off value of 0.70. The convergent validity was also analyzed through two criteria's stated by Fornell and Larcker (1981). First the indicator loadings were all above 0.70 and significant, and also the average variance extracted (AVE) of each construct exceeded the variance and were above 0.50. It is also observed that all 35 items loaded well above 0.50 which is an evidence for convergent validity (Nunally, 1978). Therefore, all the conditions were met for convergent validity

Table III. Reliability and Convergent Validity

Constructs	AVE	Composite reliability	R square	Cronbach's Alpha
Leadership	0.77	0.94	-	0.92
Continuous Improvement	0.75	0.93	-	0.91
Job Satisfaction	0.77	0.94	0.87	0.92
Management by Fact	0.80	0.92	-	0.87
OCB	0.65	0.90	0.28	0.86
People Based Management	0.84	0.94	-	0.90
All work is Process	0.79	0.95	-	0.94
Teamwork	0.83	0.93	-	0.90

4.2.Path coefficients and predictive ability

The correlations between all the variables in the study were significant and above 0.50. Moreover, job satisfaction was highly correlated with organizational citizenship behavior (r=0.80).The PLS analysis results (Table IV) indicated that two of the hypotheses had a negative effect (Hpothesis2 and hypothesis7), however the rest of the hypotheses were supported. Thus, the fit of the overall model is good. Job satisfaction increases organizational citizenship behavior (t=5.42). Continuous improvement (t=2.40), management by fact (t=2.09), people based management (t=2.37) and all work is process (t=1.78) increase job satisfaction. However, leadership (t=0.82), and teamwork (t=0.53) has non-significant relationship with job satisfaction in this study.

Table IV. Summary of the Hypothesis Test Result

Association	Hypothesis	Casual Path	Path Coefficient	t-value	Supported
OCB	H1	JS -> OB	0.53	5.42	Yes
J. Satisfaction	H2	LED -> JS	0.06	0.82	No
	H3	CI -> JS	0.20	2.40	Yes
	H4	MF -> JS	0.23	2.09	Yes
	H5	PM -> JS	0.23	2.37	Yes
	H6	AP -> JS	0.20	1.78	Yes
	H7	T -> JS	0.05	0.53	No

5.DISCUSSION

The results of this empirical study clearly present the effects of Business Excellence practices in the hospitality sector. The empirical findings also demonstrate that leadership, management by fact, processes, teamwork, employee orientation and continuing improvement are significantly and positively associated with job satisfaction.

The empirical research findings are consistent with prior empirical studies that employee satisfaction is positively related to OCB (organizational citizenship behavior), even though the sector is very challenging; salaries are not so high, the tasks require long hours working, turnover is very high and the sample consist of frontline employees in this study.

While several prior researches in the service quality management have examined the relationship between satisfaction and loyalty (Chang et al, 2010; Jun et al., 2006) in the HRM perspective, this study is the first to examine the role of Business Excellence practices in the hotel sector, by directly taking into consideration frontline employees in Tehran, Iran. The empirical results indicate both; BE practices acts indirectly on OCB through the mediating effect of job satisfaction; and leadership and teamwork have insignificant effect on employee satisfaction. Those results show that the hypotheses are supported except for two.

6.IMPLICATIONS

6.1.Theoretical implications

The outcomes of this empirical study are important to the quality literature on hotel quality management and hospitality organizational studies plus these links have never been studied previously. Since employee satisfaction significantly affects their organizational citizenship behavior within a hospitality organization, it may be an opportunity for hospitality managers to focus and increase the frontline employees or even they may be able to increase the overall employee satisfaction levels. The studied Business Excellence practices are proved to be significantly affective in frontline employee satisfaction. Therefore, the tourism and hospitality authorities can improve employee job satisfaction through Business Excellence practices.

6.2.Practical implications

The results indicate a positive significant relationship between job satisfaction and OCB in the hotels of the capital city of Iran, these results are in line with the findings of Allen and Rush, (1998), Bateman and Organ, (1983), Bishop et al. (2000) and Murphy et al. (2002), who also found positive relationships between the job satisfaction and OCB.

The research finding reported that the relationship between leadership and job satisfaction is not significant. However the other researchers found significant relationships between them (Chang et al, 2010). As the leaders take care to help and support the employees and they are concerned about the needs of their employees. The reason for having a non-significant result might be lack of trust and poor relationships between the Frontline employees and managers at the hotels of Iran.

The results indicate a positive significant relationship between continuous improvement and job satisfaction of frontline employees working in the hotel industry of Iran. Since the tourism activities in the hospitality industry are a never-ending journey, the continuous improvement should be in place, and the results indicate that the frontline employees in Iranian hotels are very good at learning from their mistakes and encouraging themselves by the feedbacks and corrective actions in the hotel industry. In addition this influences the job satisfaction level of the frontline employees which has resulted in a significant positive relationship.

Management by fact is also one other business excellence practice that has a significant positive relationship with the job satisfaction of the employees working in Iranian hotels. This indicates that the strong influence that has been made on the process management that led to the mapped, analyzed and streamlined work processes has highly influenced the job satisfaction level of the employees working in the hotel.

People base management is another business excellence practice that has a significant positive relationship with job satisfaction. The necessary resources and the reward system that is used by the managers of the hotels in the Iranian hotels have highly influenced the job satisfaction level of the employees, as this is in line with what Kanji (2008) mentioned in defining the people base management he added that it has been researched that they are highly correlated. The employees perform better if they like the environment that they work in.

All work is process which is having a better vision about the processes in the hotel is the main issue that people base management is based on, which has a significant positive relationship with job satisfaction of the frontline employees working in the hotel. This is the result of having a programmed everyday processes that has identified the customer requirements and also expectations and they are applied by all the frontline employees in the Iranian hotel industry.

Although teamwork had a significant effect on employee satisfaction in many researches (Valle et al., 2001; Chang et al, 2010), in this study the results indicate that teamwork was not supported in the hospitality industry of Iran. The non-significant relationship between job satisfaction and teamwork could result from lack of composition of the teams, the low level of trust between the employees working in the hotel industry in Iran. Various studies points out that lack of training in group work in the organizations might result in having negative results of this relationship.

To sum up it is worth mentioning that the problem of Iranian hotels with respect to their frontline employees' responses to the relationships between business excellence practices and job satisfaction that led to organizational citizenship behavior was lack of good leadership as well as good relationships between the frontline employees of the hotels and the leadership of the hotels, which might have been the result of lack of trust in today's world. The other dimension that was not significantly related to the job satisfaction of the frontline employees was the teamwork structure; this might also result from lack of trust between the employees of hotels, as it has been known to be the key mechanism in the involvement of employees in the quality improvement of the hotel. In addition, teamwork was one of the main issues that helped the employees in dealing with the complex problems that need multidisciplinary approaches (Kanji, 2008).

7.LIMITATIONS AND FUTURE RESEARCH

The limitations of the study are concerned with finding the up-to-date statistics for Iran, since the country has restrictions. The second limitation was being very hard to gather the data from the frontline employees of the hotels, since they were very busy and had no time, or they were not confident enough to answer the questions properly. The future research might take other dimensions and find their relationships with the business excellence practices in the hotel industry of Tehran, or in other industries, or worldwide.

There are serious implications in our findings for politicians and policy makers, the hotel managers are responsible for leading the hotel as well as achieving their objectives. The messages for formally chosen leaders are endorsing their role in a way that increases engagements no matter what their position in the hotel is. An engaging leader may be identified as someone who encourages and agrees to develop the hotel, and who is characterized by a culture based on honesty, openness and intelligibility, and the genuine valuing of others and of their contributions. Engaging leadership enables hotels to cope with change, and also to be practical in shaping their future. They should always be guided by ethical standards and the desire to co-create and co-own ways of working with others towards a shared vision.

Subsequently, hotel managers should also support, lead, coach and make efficient teamwork and give sufficient amount of empowerment to enhance participation and provide better training and education to back up these improvement efforts. Hotels need to look at strategies for improving performance in the light of increasingly competitive environment. Successful hotels are those which are open to internal change and persuade a culture that encourages innovation and market responsiveness. Team working is a strategy that has the potential to improve the performance of individuals and organizations in terms of co-operation and high performance, but it needs to be taken care of over time in order to obtain more competitive and innovative teams with higher quality, delivery precision and productivity. Hotel managers need to have the vision to introduce team working, the sensitivity to promote it and the courage to permit teams to play a proper part in decision making.

REFERENCES

1. Allen, T. D., & Rush, M. C. (1998), "The Effects of Organizational Citizenship Behavior on Performance Judgements: A Field Study and Laboratory Experiment", Journal of Applied Psychology, 83(2), 247-260
2. Arasli, H. (2012). Towards Business Excellence in the Hospitality Industry: A Case for 3- 4-, and 5-star hotels in Iran. Total Quality Management Journal, 23(5), 573-590.
3. Arasli, H. (2002). Diagnosing Whether Northern Cyprus Hotels are ready for TQM: An Empirical Analysis', Total Quality Management, 13(3), 347-364.
4. Arasli, H. & Baradarani, S. (2010). Toward Business Excellence in Hospitality Industry in Iran: A Case for 3, 4, 5 Star Hotels. The Second Canadian Quality Congress-Toronto (August).
5. Bateman, T.S. & Organ, D.W. (1983), "Job Satisfaction and the good Soldier: The Relationship between Affect and Employee Citizenship", Academy of Management Journal, 26,587-595.
6. Bishop, J.W. & Scott, K.D. (2000), "An Examination of Organizational and Team Commitment in a SelfDirected Team Environment", Journal of Applied Psychology, 85(3),439-450.
7. Camison, C., Flor, M., Cruz, S. & Kuster, I. (1996). Quality Practices and Perceptions of Valencian Hospitality Enterprises: An Empirical Analysis. International Journal of Quality and Reliability Management, 13(7), 79-93.
8. Chang, C.C., Chiu, C.M. & Chen, C.A. (2010). The Effect of TQM Practices on Employee Satisfaction and Loyalty in Government. Total Quality Management Journal, 21(12), 1299-1314.
9. Dale, B. & Cooper, C. (1994). Introducing TQM: The Role of Senior Management. Management Decision, 32, 20-26.
10. Denison, R.D., Hart, S.L. & Kahn, J.A. (1996), "From Chimneys to Cross-Functional Teams: Developing and Validating a Diagnistic Model", Academy of Management Journal, 39(4),1005-1023.
11. Drucker, P. (1964). On the Profession of Management. A Harvard Business Review Book.

12. Feinstein, H. & Vondrasek, D. (2006), "A Study of Relationship between Job Satisfaction and Organizational Commitment among Restaurant Employees," 2006. http://hotel.unlv.edu/pdf/jobSatisfaction.pdf

13. Fornell, C., & Larcker, D. (1981). Evaluating Structural Equation Models with Unobservable Variables and Measurement Error. Journal of Marketing Research, 18, 39-50.

14. Furnham, A., Petrides, K.V., Jackson, C.J. & Cotter, T. (2002), "Do Personality Factors Predict Job Satisfaction?", Personality and Individual Differences, 33(8),1325-1342.

15. Gillet, S. R., & Morda, R. (2003), Effective Leadership in Tourism and Hospitality Organizations in the 21st Century, In S. Kusluvan, Managing Employee Attitudes and Behaviors in the Tourism and Hospitality Industry (pp. 599-630). New York: Nova Science Publishers.

16. Griffin, M.A., Patterson, M.G. and West, M.A. (2001), "Job Satisfaction and Teamwork: The Role of Supervisor Support", Journal of Organizational Behavior, 22(5), 537–550

17. Hansmann, K.W., & Ringle, C.M. (2004). Smart PLS Manual. Hamburg: University of Hamburg.

18. Hartline, M.D., & Ferrell, O.C. (1996). The management of customer-contact service employees: An empirical investigation. Journal of Marketing, 60, 52–70.

19. Heslop, Heslop, P., Smith, G. D., Metcalfe, C., Macleod, J., & Hart, C. (2002), "Change in Job Satisfaction and its Association with Self-Reported Stress, Cardiovascular Risk Factors and Mortality", Social Science & Medicine, 54,1589-1599.

20. Herzberg, F., Mausner, B. & Snyderman, B. (1959), The Motivation to Work. New York:Wiley.

21. Hulland, J. (1999). Use of Partial Least Squares (PLS) in Strategic Management Research: A Review of Four Recent Studies. [illegible], *)(2), 195-204.

22. IRAN Cultural Heritage and Tourism Organization. (2006). Iran Tourism Information (ICHTO) from: http://iranmiras.ir/en_site/Home_E.asp.

23. IRAN Daily. (2006). Iran Hotel Information from: http://www.irandaily.com/1383/2239/html/focus.htm.

24. Jun, M., Cai, S., & Shin, H. (2006). TQM Practice in Maquiladora: Antecedents of Employee Satisfaction and Loyalty. Journal of Operations Management, 24(6), 791–812.

25. Juran, J. M. & Gryna, F. M. Jr (Eds) (1988). Juran's Quality Control Handbook, 4th ed., McGraw-Hill, New York, NY.

26. Kanji, G. K. (2002). Measuring Business Excellence. Routledge. • Kanji, G. (2008). Architecture of Business Excellence in the Public and Service Sector. Total Quality Management Journal, 19(4), 399-415.

27. Koc, E. (2006). Total Quality Management and Business Excellence in Services: The Implication of AllInclusive Pricing System on Internal and External Customer Satisfaction in the Turkish Tourism Market', Total Quality Management and Business Excellence, 17(7), 857-877.

28. Locke EA. (1969), What is job satisfaction? Organ Behav Hum Perform. 4(November),309–36.

Chapter 14

Motivating Your Staff to Provide Outstanding Service

Michael C. Sturman, Robert Ford

Cornell University, mcs5@cornell.edu

University of Central Florida

ABSTRACT

Your service environment is superb, your operations are efficient, your concept reflects a clear idea of the customers' demands, and your back-of-the-house delivery system is flawless—so far, so good. Now it's up to that front-line employee you've worked so hard to hire and train to deliver the service you've promised, because the guest contact employee usually is the key to having a satisfied guest. Say that there's a problem at the front desk, and a guest's room is out of order or not acceptable. This is the moment of truth. If the frontdesk associate is rushed or dealing with a heavy check-in, he or she may not be able to respond to the guest's problem or may be too casual about the problem. Suddenly, all your careful planning is for naught. If the guest feels shabbily treated, the guest will also be angry at your hotel, since the employee represents the hotel. It makes no difference whether the guest's reaction is fair or appropriate. Since quality and value are defined by the guest, the employee who provides the guest experience must be not only well trained but highly motivated to meet the guest's quality and value expectations and to do so consistently. Your role as a manager responsible for providing an exceptional service experience, and specifically for preparing your staff to deliver such service, is vital. In this chapter, we argue that motivating your employees is as critical for excellent service as is training them.

Keywords: Cornell University School of Hotel Administration, hospitality management, hospitality industry, employee motivation, customer service, guest experience

Your service environment is superb, your operations are efficient, your concept reflects a clear idea of the customers' demands, and your back-of-the-house delivery system is flawless—so far, so good. Now it's up to that front-line employee you've worked so hard to hire and train to deliver the service you've promised, because the guest contact employee usually is the key to having a satisfied guest. Say that there's a problem at the front desk, and a guest's room is out of order or not acceptable. This is the moment of truth. If the front-desk associate is rushed or dealing with a heavy check-in, he or she may not be able to respond to the guest's problem or may be too casual about the problem. Suddenly, all your careful planning is for naught. If the guest feels shabbily treated, the guest will also be angry at your hotel, since the employee represents the hotel. It makes no difference whether the guest's reaction is fair or appropriate. Since quality and value are defined by the guest, the employee who provides the guest experience must be not only well trained but highly motivated to meet the guest's quality and value expectations and to do so consistently. Your role as a manager responsible for providing an exceptional service experience, and specifically for preparing your staff to deliver such service, is vital. In this chapter, we argue that motivating your employees is as critical for excellent service as is training them.

MOTIVATING EMPLOYEES

Let's look at the way in which a motivated front-desk clerk handled an actual room reservation problem. A family was checking in at the Hyatt Grand Cypress on a busy night. The hotel was full, the family's reservation had not been properly handled, and the husband, wife, and three tired children were upset. The front-desk employee assessed the situation and acted promptly. Rather than force the family to hang out in the lobby until things were straightened out, she took some quarters out of the petty cash drawer and gave them to the kids to go and play the video games and gave the parents chits for a drink in the lobby bar. That gave her the necessary time to find a manager to straighten out the problem. The parents were happy, the kids were happy, and the front-desk person had defused a tense situation.

In contrast, you may have heard about the experience of Tom Farmer and Shane Atchison, whose situation eventually went viral on the Internet. They had a confirmed reservation at a hotel, and the room had supposedly been guaranteed by credit card for late arrival. Yet when they arrived a 2:00 AM, they were refused rooms. Tom and Shane publicized their service experience by posting an open complaint to the hotel, called "Yours Is a Very Bad Hotel."1 They report that "Mike the Night Clerk" was "deeply unapologetic," "had done nothing about finding [them] accommodation elsewhere," and when they suggested that he should have lined up other rooms in advance, Mike the Night Clerk "bristled!" Explaining why the customers were wrong to be upset that their "guaranteed" room was not saved for them, Mike told them, "I have nothing to apologize to you for." After the posting, representatives of the hotel did apologize and reportedly improved their training and overbooking policies, but the damage was done. A Google search in December 2010 on the phrase "Yours Is a Very Bad Hotel" yielded 15,200 results!

Both of the above situations had the same essential cause: a failed reservation. The key here is that the employees' actions brought markedly different results. The Hyatt's desk clerk was motivated to keep the customers satisfied despite a problem that was not of her making, whereas Mike the Night Clerk apparently could not have cared less. As you read this chapter, consider how your employees might respond to a common situation like this one. Your challenge is to discover what motivates your employees not only to do their jobs efficiently and competently but also to go the extra mile for your guests.

In this chapter, we discuss various ideas that will help you motivate your staff. We have divided these ideas into financial approaches and nonfinancial methods—all aimed at helping you motivate your staff to offer the best possible guest experience even when things don't go right. We have all seen that it's impossible to predefine policy and procedures for all guest situations. Something unexpected always comes up. Thus, employees should know that they are encouraged, expected, and trusted to handle all the many and varied situations that come up in the guest service areas for which they are responsible. If they were properly selected and trained in the first place, management must make it possible for them do their jobs with responsibility, skill, enthusiasm, and fun. The purpose of this chapter is to consider the many ways you can accomplish this.

The motivation, the drive, or compelling force that energizes people to do what they do comes from either inside the employees (intrinsic motivation) or from the environment (extrinsic motivation). The best organizations create an environment and policies that enhance both forms of motivation. It's critical to note right at the start that money—important though it may be—is not your employees' only motivating factor. You can motivate employees not only by offering financial rewards, but also providing nonfinancial rewards, or even changing the nature of the job itself. Decades of economic and psychological research have shown that motivation is influenced by numerous factors, including the organization's environment, organizational culture, individual characteristics, comparison across individuals, and the structure and form of existing incentives. Unfortunately, it is impossible for us to offer you simplistic prescriptions on how to increase your employees' motivation, but you do have a wide variety of tools at your disposal to create and sustain an environment where employees want to go the extra mile.

FINANCIAL REWARDS

Let's start with the most common motivator—money. Your employees receive direct compensation (wages or salary). You probably also offer indirect compensation, which puts money in your employees' pockets by giving them services and benefits, such as deferred compensation and health insurance. You may also have at-risk pay, where rewards (if they are given at all) depend on various factors, like individual, team, or organizational performance.

Psychological research is filled with different theories and studies of motivation, many of which are applicable to understanding financial incentives. We won't describe all of these ideas, but let's look at four factors that researchers have identified for why money motivates your workers—or fails to do so. Those factors are individual needs, the power of reinforcement, the expectancy of getting rewards, and the fairness of the rewards.

The Power of Individuals' Needs

To meet employees' living needs, the most obvious inducement is money. Most of your employees are working because they need to earn a living—even if they work in your hotel or restaurant for other, nonmonetary reasons. Motivational speakers like to talk about higher-order needs like social, recognition, or achievement needs, but none of those count if a person is concerned about how they are going to get their next meal, pay their rent, or buy warm clothes in the winter.

You must pay your employees enough to meet their survival needs before they can effectively use incentives or rewards to help satisfy their other needs.2 While you may not be in the financial position to suddenly increase wages or offer health benefits, you are not powerless either. There are other valued incentives to help them, such as scheduling enough hours, working around child or elder care responsibilities, and helping employees find affordable health care for their families. Listening to employees' needs and preferences (as described in Chapter 4) often reveals simple available solutions. Such an approach can go a long way to satisfy employees' needs.

The Power of Positive Reinforcement

The idea of positive reinforcement is a well-accepted psychological principle. If you reward desired behavior and don't reward undesired behavior, you will get more of the desired behaviors and fewer of the undesired ones. While the idea seems straightforward, successfully using positive reinforcement in the workplace is challenging. Too often, time pressures compel you to focus on the annoying "squeaky wheel" rather than the quietly effective employee. Rewarding the wrong behavior is as big a mistake as not rewarding the right behavior

Here's an example of how improper reinforcement unintentionally changed the nature of a greeter position for one restaurant. The manager initially told the newly hired employee explicitly that his primary responsibility was to greet and welcome guests. As time went on, however, to keep the employee busy when guests were not entering the restaurant, the manager added responsibilities to the position—such as checking periodically to make sure there were enough trays or making sure that the butter dish was always full. The greeter quickly realized that the manager never complimented him for properly greeting the guests, nor did the manager ever say anything to him when he missed a guest because he was too busy with his other duties. But if he ever let the buffet line run out of trays or butter, he was strongly reprimanded. Therefore, by reinforcing certain employee actions, the manager had redefined the job. The manager made her real priorities clear, the employee adjusted his actions accordingly, and the restaurant no longer reliably offered a greeting to guests.

The Power of Goals and Expectancy

Another key way to motivate employees is by setting challenging yet still attainable goals. Research on goal setting has repeatedly shown that setting specific and challenging goals leads to higher levels of task performance than do easy or vague goals, or telling people to "do their best." If you couple challenging goals with compensation, you have a powerful force to motivate exceptional employee performance. Of course, as shown by the case of the buffet greeter, rewarding the wrong behavior can sabotage motivation. Goals could be too easy or too challenging, may induce unethical behavior or inappropriate risk taking, or may motivate behaviors leading to rewarded tasks at the expense of other, equally important, but unrewarded tasks.3

Harrah's casino uses a combination of goal setting and future rewards to motivate both individual and team behavior. Sales staff employed in Las Vegas are rewarded for making sales goals set for the individual person, the hotel, and Harrah's group of hotels in the Las Vegas area. Managers work with the teams to make sure the goals are challenging but attainable. Successful performance is rewarded with both bonus checks and merchandise that the employees can pick out (such as Weber grills and vacation packages). These goals extend beyond simply sales numbers. All employees can achieve bonuses based on a number of critical metrics, including customer service scores. Setting these goals, rewarding them, and using goals that are clearly connected to the company's strategy of excellent customer service allows Harrah's to get highly motivated staff and exceptional performance

The Power of Equity

Reinforcement, goal setting, and expectancy assume that people make decisions based purely on their own circumstances, but people are social creatures. We see what others do and get, compare ourselves against others, and decide what is fair or unfair, just or unjust. If someone thinks that other employees were given more rewards for similar efforts, this will create feelings of injustice. Feelings of being underpaid have been linked to lower motivation, greater absenteeism, higher turnover, and even increased theft. We hasten to add that employees' feelings of lack of equity involve being comparatively underpaid—few people admit that they are overpaid and feel bad about it

Individuals will compare against others, both inside and outside of the company. Your employees will compare themselves to their coworkers, to subordinates, and even to the boss, all the time evaluating whether pay differences are (in their eyes) just or unjust. Your employees will also try to figure out what the "market" is paying for their job, talk to friends who are doing similar jobs at other organizations, look at other jobs they might be able to take, or even consider other careers they might pursue. Making this even more confusing, people will often form judgments of fairness based on their perceptions of their own worth or perceptions of others' pay—neither of which may be accurate.4

Most know how important fairness is in compensation systems. Indeed, for the process of determining Fortunes "100 Best Companies to Work For" list, compensation fairness is a key criterion.

Types of Performance-Based Financial Rewards

With those principles in mind, let's look at possible performance-based financial rewards. Your challenge is to discover what employees consider equitable and appropriate and then provide rewards that fit those criteria, while also being fiscally prudent. To make rewarding good performance even harder, employees have changing expectations, moods, and valuations of the employment relationship.

- Your first step is to identify the rewards that employees desire. Most employees begin new jobs with energy and enthusiasm. They want to do well. So your motivation program should build on this good start by offering encouragement and help. By giving your employees desired rewards for performance, you can keep people energized, enthusiastic, and working hard. Fortunately, you can choose among many different sorts of financial rewards to do this. We've summarized these rewards in Table 11.1.
- Let's look at some of these financial incentives, many of which you may already be using: Annual performance bonuses reinforce recognition of individual performance while helping people focus on achieving individual performance goals and recognition.
- Group financial incentives reinforce social needs by rewarding team success and thereby focusing employees on working in a group to achieve common goals.
- Individual tips can encourage high customer service, while pooled tips can encourage servers to work together.
- Ownership in the company (through stock) helps employees realize how their personal success depends on the company's overall success. This can satisfy social needs (i.e., when the company succeeds, our social group succeeds) and achievement needs (i.e., when the company achieves, I also feel a sense of achievement).

Depending on how a financial reward is presented, it can help satisfy recognition needs as well as financial needs. So different pay methods can focus your employees on achieving high performance. As a caution, though, just as the manager's reinforcement changed the job of the buffet greeter, financial incentives can change behaviors, and so you should be sure that the behaviors you are rewarding are the ones you truly want to see in your employees

Financial rewards can be a powerful motivator that can reinforce your efforts to build a culture of service excellence. Your employee motivation program can be strengthened even further with nonfinancial rewards, as we explain next.

Nonfinancial Rewards

Many of the reasons why people behave as they do have nothing to do with money (e.g., bringing up children, helping a stranger, volunteering in the community, caring for parents). Consequently, nonfinancial rewards like recognition, praise, personal growth, and an enjoyable working environment (which cost you little or nothing) can produce results that money could never buy. Let's look at the most common nonfinancial rewards.

Recognition Programs

The purpose of a recognition program is to say "thank you" or "well done." Nearly 90 percent of companies offer some form of recognition program. You can recognize your employees for length of service, exceptional performance, sales performance, innovative suggestions, safety, or attendance, as well as celebrating their retirement. The rewards that go along with these recognitions include plaques or certificates, company logo merchandise, gift certificates, jewelry, office accessories, household items, recreational items, and electronics—and even a night or two as a guest of your hotel.5 These awards are designed to reinforce employee behaviors that support and sustain the company's mission. Awards like these can clearly demonstrate that you desire and appreciate your employees and their work efforts

A key advantage of these recognitions is that they can create a lasting memorable experience. Not only does the employee have the award itself, but that person has feelings of pride from receiving a plaque, lasting feelings of appreciation or success by having a trophy, or fond memories stirred by photos from a special vacation. Often, the emotional connection of a physical recognition is much stronger than what you'd get from receiving an equivalent amount of money. For example, a trophy that announces an employee's outstanding service may cost only $20, but the emotional impact of receiving a trophy is far greater than that of simply receiving a $20 bill. The trophy presentation can be something that the employee never forgets.

Fairmont Hotels developed a recognition program designed to reinforce their service culture. The program features different award levels, such as "Memory Maker" for an outstanding demonstration of thoughtfulness and creativity, and "Star of the Month" for those who consistently demonstrate superior performance.6 Employees at all levels are nominated by peers. Thus, the program helps support Fairmont's culture of appreciation and recognition for everyone. Award winners receive gift certificates in small denominations (typically $10 to $50) that can be redeemed with thousands of merchants from around the world. Surveys of Fairmont's employees after the program was implemented revealed that employees felt more recognized for a job well done, felt the best performers were more likely to receive recognition, and felt that they were being recognized in a way that was personally meaningful. Surveys also showed that employees' level of engagement increased significantly since the program's inception.

Making the Job Pleasurable

Sometimes the work itself can be its own reward. Walt Disney said, "You don't work for a dollar—you work to create and have fun."7 Chili's Norman Brinker said, "If you have fun at what you do, you'll never work a day in your life. Make work like play—and play like hell."8 Employees who like what they do continue to do it. Not surprisingly, the reverse is also true; dozens of research papers have shown that low employee satisfaction is a strong predictor of employee turnover.9 The best hotel companies regularly survey their employees to assess their job satisfaction and other attitudes, such as employee relationships with their bosses, their feelings about their pay, and opinions about working conditions.10 You might consider asking your employees what part of their job holds them back from doing their best, so that you can reconfigure their jobs to focus on the parts that they like best—and therefore perform best. What one employee dislikes, another may find just ideal.

Making the Job Interesting

Obviously, you want to hire people who already want to do the job you hired them to do and are able to perform their job responsibilities. You can make sure of this through an effective recruitment and selection process. But once you have hired qualified people, you want to make sure they remain interested in performing the specifics of their job at the level you need them to. You can make most jobs interesting by using one or more of the following four elements:

Give employees the ongoing opportunity to learn and grow, both personally and professionally. People often want to feel a growing sense of professional mastery in their area of interest, but jobs with little to no variation can lead to burnout. Employees may have the ability to perform well, but may have done the same routine task so many times that they simply don't feel like doing the job as well as they can. Job rotation may be effective for some employees in this case, but others want to continue to perform jobs for which they comfortably feel they are qualified. So you might also consider ways to develop new interactions, new tasks, and new challenges. 2. Workers expect to be treated with respect. As employees learn and grow, you may show this by increasing employees' personal responsibility. Increased autonomy can be a sign of organizational respect for their knowledge, skill, and capabilities. In more dayto-day ways, employees need to be treated with integrity (as discussed in Chapter 31). If workers are not treated with respect, the only factor keeping them connected to the company will be their paychecks. If that is the case, once another company provides an equal or better offer, those employees (and all that you invested in them) may walk out your door.

3 Make each job important by making clear its purpose in the organization. If employees think that their job performance really makes a difference in the organization's ability to provide an excellent experience to the customer, then they will feel good about performing their roles well. They will have pride in their job and their organization. Finding ways to show employees that what they do makes a difference to an important operation is a key motivational skill of managers of successful hospitality firms.

4.Arrange to satisfy individuals' social needs as part of the job. Many people want to be part of a good work group, or more generally have relationships with other people where they work. This is particularly true in the hospitality industry. Well-designed work groups can be helpful in managing employee direction and behavior in the workplace. Indeed, as studies at the Hawthorne Plant of the Western Electric Company showed many years ago, the sense of belonging or not belonging greatly influences what people will or won't do in the workplace.11 The managerial focus here should be to work in harmony with the group to support each employee's effort to achieve the group goal, which will help achieve your hotel's goal.

Groups may also offer members the opportunity to achieve something greater than themselves by being a part of the group. Membership in organizations with strong corporate cultures and whose purposes are respected by the society as a whole is valuable to both groups and individuals. Asking a group to help accomplish the valued, respected organizational purpose becomes a powerful motivational tool and a primary means for keeping the individual and the group positively involved in your operation.

MOTIVATION AND CORPORATE SOCIAL RESPONSIBILITY

JoAnne Kruse, Founder@www.HC-partners. Net

In these challenging economic times, most human resources (HR) organizations are struggling with shrinking budget allocations for services that were historically considered standard, like training, recognition, and incentives. But even in a labor market flush with available talent, the concepts of retention and engagement are of critical importance. At this writing, the unemployment rate hovers around 10 percent in the United States, but how does an employer keep those employees still on-board engaged and motivated to perform without access to the traditional motivators of compensation and career mobility? Adding even more to complicate the equation is the social tone that resists any level of conspicuous consumption. This results in a workplace environment that encourages a low-cost approach to motivation. Together, these dynamics create an ongoing pressure for employers to deliver on true productivity gains while retaining talent when the market does (eventually) pick back up.

Even before the Great Recession, sectors like travel and hospitality were dealing with significant volume declines. Barely able to recover from the impact of 9/11, the travel sector then faced one global crisis after another, including the 2004 tsunami, terrorist bombings in Bali and London, and Severe Acute Respiratory Syndrome (SARS). As the chief HR executive in the travel sector with Travelport Ltd., a travel technology company, I saw the industry enter the millennium dealing with rapid erosion of the top line and a heightened focus on cost containment. At the same time, the employment market was strong, making retention and talent acquisition challenging. Part of our strategy included developing a corporate social responsibility (CSR) strategy to differentiate ourselves with customers and employees. We believed that employee productivity could be improved through higher levels of employee engagement. Applying Corporate Leadership Council research on engagement, we developed a broad array of programs to build employee engagement to drive effort and results. Our CSR initiative was one such program, combining our internal focus on engagement with good business and customer initiatives

The tragic, deadly Indian Ocean tsunami of 2004 acted as a call to action for Travelport management, and the impetus to develop a broader CSR program that included "voluntourism," charitable giving, and environmentalism. The program had three core goals: engage employees around the world in local volunteer and giving programs; create and deliver against a green strategy for our hundreds of facilities located globally; and align our program investments with our business development and customer interests. The horrific destruction of the tsunami in Thailand provided a critical need, and our initial investment centered on a fund-raising effort to rebuild a destroyed orphanage in the resort area of Phuket. We designed the program with a global logo, Web-based donation site, and ongoing postings of pictures and fund-raising results on the company intranet. The program was successful, raising over $250,000 in a little over three months—with all regions participating, including those where charitable efforts are less culturally common than in the U.S. Employees rallied around the company and the commitment demonstrated to "doing the right thing," providing a bonding moment for local teams and a recommitment to the executive team's leadership. In the midst of significant uncertainty and marketplace challenges, employee satisfaction soared, and feedback from the event continued for years following. The broader CSR program then followed suit, with a commitment to "design globally, apply locally," further reinforcing the company's organizational structure and approach to product and service delivery. Further, the opportunity to deliver a positive message to customers proved a much-needed boost for our sales and service teams.

The CSR program became a living example of how the Travelport culture and values were demonstrated—to our employees and customers—around the world. By "walking the talk" we were able to reengage a change-weary employee base and motivate employees and customers around the world by delivering a sense of pride in the company.

The CSR program became a living example of how the Travelport culture and values were demonstrated—to our employees and customers—around the world. By "walking the talk" we were able to reengage a change-weary employee base and motivate employees and customers around the world by delivering a sense of pride in the company.

Finally, let's look at the interaction of the employee with the job itself Many people are powerfully motivated intrinsically, or internally. Unattached to specific rewards of any kind, intrinsic motivation is caused by the nature of the person and their environment. One way to take advantage of this is to hire people motivated to perform exceptional service. Another way to foster intrinsic motivation is by structuring the job and work environment to build on employees' personal motivation.

LEADERSHIP

It is impossible to overstate the importance of your role as a leader. Leaders must also be able to identify and provide those rewards that the individual employee wants from working in your hotel or restaurant. Effective leadership involves inspiration, challenge, and a strategic vision. Creating a shared purpose among employees can effectively foster motivation.12 A great leader creates the conditions that promote employee engagement.

Charles A. Conine, SPHR, Cornell 1973

Detroit, Michigan, 1977. The morning sun was reflected in prisms of light off the newly opened, 73-story Detroit Plaza Hotel at the Renaissance Center, the glass centerpiece of and home to what was billed as the Motor City's resurgence. Nearly 2,500 Detroiters had been hired to staff this new

Westin hotel. Many, due to Detroit's riots, chronic unemployment, and systemic poverty, had never worked a day in their adult lives before being put on the payroll at the "Ren Cen." My task preopening had been to shepherd 12,000 job applicants, sifting through their midst to identify those whose attitude was at least good— not an easy job in a bruised and battered city where "hope" was far from the most common of four-letter words.

Now, with the hotel finally open, as director of personnel, I watched as right away, immovable and opposing forces assembled. On one side were 2,250 union members led by a feisty union business representative. On the other side was Westin's large and as yet unproven management team. "Battles royal" raged for several months, with employees, sometimes at the union's behest, often camping outside my door three deep, hankering to get a crack at convincing me why their boss was guilty of racism and why the employees shouldn't be picked on. Management felt rather picked on, too. I felt engulfed, without a way out.

Finally, in desperation I met with the union business rep, who, to her everlasting credit, suggested we create a shop steward's council, a place for the hotel's departmental union representatives to go and air issues without fear of retribution. We soon found ourselves there, the opposing forces eyeing one another warily before the floodgates opened and out poured dozens of complaints, simmering issues, and stories of mistreatment.

Over the course of three successive steward meetings, however, as issues were heard and, in most cases, quickly addressed, the number and ferocity of complaints gradually subsided. We had turned a corner, discovering what really worked: the knowledge that when management and subordinates care equally about respect and working conditions, mutual animus recedes as mutual trust ascends

Knowing that your concerns will be heard and taken seriously can be a powerful motivator. Indeed, as the nascent truce took hold, its calm spreading as everyone's motivation to work together improved, the "climate" inside the hotel, in that crucible I'll forever recall as simply "the Spring of 1977," became downright balmy.

Empowering Employees

Most hospitality operations function with employee teams, which need to grow and develop just as individuals do. One great asset that a team provides to its members is the opportunity to grow within the group setting. But the organization must provide additional opportunities for its members to satisfy this important need. The most widely discussed strategy for doing so is empowerment

Empowerment is the assignment of appropriate decision-making responsibility to the individual.13 Empowerment is broader than the traditional concepts of delegation, decentralization, and participatory management. Empowerment stretches decision responsibility to include the entire job and ensures that the performance of that job fits with the organizational mission. The empowered server can personalize the service experience as it relates to each guest's expectations and can take whatever steps are necessary to prevent or recover from any service failures.

The danger in empowerment is that some organizations promise employees decision input without giving them any real power or authority. Employee empowerment must not only ensure that effective decisions are made by the right employees, but it must provide a mechanism by which responsibility for job-related decisions is vested in either individuals or teams.14 Empowerment also means that management is willing to share relevant information about and control over factors that impinge on effective job performance. You are not obliged to ratify every employee decision. If the employees make a decision contrary to your hotel's policies, further discussion is required, but you cannot second-guess every employee decision.

Hyatt Hotels, for example, felt that they should provide better hospitality to their most loyal customers. So Hyatt began a program that empowered employees to perform what they called "random acts of generosity." One member in the loyalty program might suddenly find that his bar tab was paid, another may receive a complimentary massage, or a member and her family may discover that the hotel has paid for their breakfast. As you can imagine, there's no way to dictate a set of policies that would tell employees how to surprise and delight guests. Instead, Hyatt empowered their employees to see a situation and make the right decision to provide exceptional customer experiences. The intent of the empowerment is to help fulfill the company's mission, which they define as making a difference in the lives of the people they touch—including guests and employees.

Putting the Pieces Together

Motivation is so complex that no single policy will help you motivate your employees. While you cannot ignore the reality that financial rewards play a critical role in motivating hospitality employees, you cannot also ignore the reality that money isn't everything to everyone. Recognition and feedback can also be effective tools, if you have the leadership and management skills to know when and how to provide these nonfinancial rewards effectively. The work itself can be highly motivating if designed appropriately. Yet a job cannot be designed without considering your hotel's culture, the training employees receive, the types of employees selected, and the way the service product is delivered. Putting all the pieces together in ways that lead every employee to do what they must do for organizational success is the true test of a manager's skill.

For the most part, hospitality employees look for jobs that are fair, fun, interesting, and important. They appreciate leadership by managers who can determine what each person is looking for in the employment relationship and can provide it consistently and fairly. Given what we know about motivation in the hospitality industry, the key to managing and retaining these employees is to create jobs that fulfill your employees' quest for fun, fair, interesting, and important positions; allocate rewards fairly; and provide leadership that takes the time and makes the effort to ensure that employees are appropriately treated, rewarded, respected, and recognized. If you can successfully build these elements into the job situation, employees will be motivated to work hard and follow your leadership. The trick is that everyone's definition of fair, fun, interesting, important, and appropriately managed is different. Thus, we must conclude by saying that the fundamental key to motivation is to know your employees.

NOTES

1. For example, see www.snopes.com/ business /consumer /badhotel.asp.
2. A. H. Maslow, "A Theory of Human Motivation," Psychological Review 50(4) (1943): 370-396.
3. G. P. Latham, Work Motivation: History, Theory, and Practice (Thousand Oaks, CA: Sage, 2007).
4. C. Albrecht, "Sales Compensation and the Fairness Question," Workspan 8(9) (2009): 17-20.
5. World at Work, A Survey of Members of WorldatWork and the National Association of Employee Recognition (NAER), May 2005.
6. M. Smith and D. Irvine, "The Power of Fairmont Hotels and Resorts' Strategic Employee Recognition Program," Workspan 52(8) (2009): 28-32.
7. W. Disney, Famous Quotes (Anaheim, CA: Walt Disney Theme Parks and Resorts, 1994), 36.
8. N. Brinker and D. T. Phillips, On the Brink: The Life and Leadership of Norman Blinker (Arlington, TX: Summit, 1996), 191.
9. N. P. Podsakoff, J. A. LePine, and M. A. LePine, "Differential Challenge Stressor-Hindrance Stressor Relationships with Job Attitudes, Turnover Intentions, Turnover, and Withdrawal

Behavior: A Meta-analysis," Journal of Applied Psychology 92 (2007): 438-454.

10. J. Barsky, C. Frame, and J. McDougal, "Variety of Strategies Help Improve Employee Satisfaction," Hotel and Motel Management 219(21) (2004): 8.

11. E. Mayo, Hawthorne and the Western Electric Company: The Social Problems of an Industrial Civilisation (Oxford, UK: Routledge, 1949).

12. J. C. Tinimerman, "A Systematic Approach for Making Innovation a Core Competency," Journal of Quality and Participation 31(4) (2009): 4 — 10.

13. Based on R. C. Ford and M. D. Fottler, "Empowerment: A Matter of Degree," Academy of Management Executive 9(3) (1995): 21—28.

14. L. C. Plunkett and R. Fournier, Participative Management: Implementing Empowerment (New York: Wiley, 1991), 5.

Index

A

Administration, 37, 41, 62, 74, 224, 237, 287
advantage, 2, 6, 25, 26, 28, 29, 268, 272, 293, 297

B

behavior, 7, 9, 21, 75, 78, 80, 281, 290, 291, 295
Behavioural, 195, 237

C

Citizenship Behavior, 270, 277, 283
co-alignment, 25, 26, 29, 30, 31, 33, 34, 35, 36, 39
compensation, 45, 47, 48, 51, 291, 296
Competence, 159, 160, 161, 164, 172, 175, 176

D

decision making, 18, 42, 48, 51, 90, 96, 101, 109, 110

E

Economics, 76, 79, 80, 192, 195, 221, 237, 266
Ethical climate, 2, 21
Ethics, 1, 2, 3, 4, 5, 7, 13, 14, 20, 21, 22, 23, 239
Excellence Practices, 270
exit surveys, 179, 187, 190, 191

G

guest experience, 286, 287, 288

H

Historical evolution, 42
hospitality, 25, 26, 27, 28, 29, 300

human resources, 3, 10, 25, 295

I

Innovative organizations, 225
Interface, 41, 42
Intrapreneurial, 224, 225, 227, 229, 231

J

job satisfaction, 55, 56, 57, 59, 279, 280, 281, 285, 294

M

management, 1, 2, 3, 5, 6, 7, 287, 288, 296, 298, 299

N

Novi Sad, 55, 56, 61, 62, 63, 67, 70, 72, 73, 75, 79

O

organization, 2, 8, 12, 13, 15, 225, 226, 227, 228, 229, 298

O

Organizational climate, 225, 227, 240

P

Psychology, 74, 75, 76, 78, 79, 267, 268, 269, 283, 301

R

Recruitment, 159, 167, 169, 170, 173, 176, 177, 178

S

Satisfaction, 55, 79, 179, 182, 284, 285, 301
service orientation, 55, 56, 57, 79, 273, 274
strategic human resource management, 52, 109
Strategic management, 25, 38, 75
Suggestion, 195

W

Workplace, 159, 175, 252, 266